Springer Theses

Recognizing Outstanding Ph.D. Research

Aims and Scope

The series "Springer Theses" brings together a selection of the very best Ph.D. theses from around the world and across the physical sciences. Nominated and endorsed by two recognized specialists, each published volume has been selected for its scientific excellence and the high impact of its contents for the pertinent field of research. For greater accessibility to non-specialists, the published versions include an extended introduction, as well as a foreword by the student's supervisor explaining the special relevance of the work for the field. As a whole, the series will provide a valuable resource both for newcomers to the research fields described, and for other scientists seeking detailed background information on special questions. Finally, it provides an accredited documentation of the valuable contributions made by today's younger generation of scientists.

Theses are accepted into the series by invited nomination only and must fulfill all of the following criteria

- They must be written in good English.
- The topic should fall within the confines of Chemistry, Physics, Earth Sciences, Engineering and related interdisciplinary fields such as Materials, Nanoscience, Chemical Engineering, Complex Systems and Biophysics.
- The work reported in the thesis must represent a significant scientific advance.
- If the thesis includes previously published material, permission to reproduce this must be gained from the respective copyright holder.
- They must have been examined and passed during the 12 months prior to nomination.
- Each thesis should include a foreword by the supervisor outlining the significance of its content.
- The theses should have a clearly defined structure including an introduction accessible to scientists not expert in that particular field.

More information about this series at http://www.springer.com/series/8790

Sidong Liu

Multimodal Neuroimaging Computing for the Characterization of Neurodegenerative Disorders

Doctoral Thesis accepted by
the University of Sydney, Sydney, Australia

Author
Dr. Sidong Liu
School of Information Technologies
The University of Sydney
Sydney, NSW
Australia

Supervisor
A/Prof. Weidong (Tom) Cai
School of Information Technologies
The University of Sydney
Sydney, NSW
Australia

ISSN 2190-5053   ISSN 2190-5061   (electronic)
Springer Theses
ISBN 978-981-10-9898-7   ISBN 978-981-10-3533-3   (eBook)
DOI 10.1007/978-981-10-3533-3

Printed on acid-free paper

This Springer imprint is published by Springer Nature
The registered company is Springer Nature Singapore Pte Ltd.
The registered company address is: 152 Beach Road, #22-06/08 Gateway East, Singapore 189721, Singapore

Supervisor's Foreword

Neuroimaging has transformed the way we study the human brain under both normal and pathological conditions. The anatomical and functional information in neuroimaging data has an important role in both brain research and clinical management of neurological and psychiatric disorders. In order to extract such information, advance our understanding of brain disorders and accelerate its translational impact, we need to develop innovative computational algorithms and methods to process and analyze these high-dimension and high-volume neuroimaging data.

Multimodal neuroimaging data, acquired from the same subject with different neuroimaging techniques or protocols, such as PET/CT, PET/MRI and MRI/DTI, enables us to explore the different brain functions and structures at the same time. However, computing the information in multimodal data is even more challenging, due to the inconsistent image temporal / spatial resolutions, contrasts, and qualities. As a result, multimodal neuroimaging computing always involves pre-processing, feature extraction, pattern recognition, and visualization techniques, varying in applications.

This book covers many aspects of brain image computing methods, and illustrates the scientific understanding of neurodegenerative disorders cohering around 4 general themes of multimodal neuroimaging computing, including neuroimaging data pre-processing, brain feature modeling, pathological pattern analysis, and translational model development. It demonstrates how multimodal neuroimaging computing techniques can be integrated and applied into neurodegenerative disease research and management, with many examples and case studies. It also contains a number of interesting extension topics, including longitudinal neuroimaging study, subject-centered analysis, and brain connectome. In all, this book introduces a series of innovative approaches and fundamental techniques in neuroimaging computing, which will greatly benefit the neuroscience researchers and neurology practitioners who are interested in medical image computing and computer-assisted interventions.

Sydney
October 2016

A/Prof. Weidong (Tom) Cai

Abstract

Neurodegenerative disorders, such as Alzheimer's Disease (AD), Parkinson's Disease (PD), Vascular Dementia (VD) and Frontotemporal Dementia (FTD), will become a global burden over the forthcoming decade due to the increase of aging populations. The characterization of neurodegenerative disorders has an important role in patient care and treatment planning, especially in the early stage of the disease, since current disease modifying agents are mainly effective before the clinical symptoms appear.

The revolutionary non-invasive neuroimaging technologies have transformed the way we study the brain, and become an essential component in the management of neurodegenerative disorders. The growth of neuroimaging studies has spurred a parallel development of image computing methods, which focus on the computational analysis of the brain images using both computer science and neuroscience techniques.

Multimodal neuroimaging enhances the neuroscience research by compensating the shortcomings of individual imaging modalities and by identifying the common findings from different imaging sources. Multimodal neuroimaging has become one of the major drivers in neurodegeneration research due to the recognition of the clinical benefits of the multimodal data and better access to the imaging devices. There is an imperative need for the development of novel multimodal neuroimaging analysis methods to address the variations in spatiotemporal resolution and merge the biophysical/biochemical information in multimodal neuroimaging data, thus enabling more accurate characterization of the complex pattern of neurodegenerative pathologies.

This study aims to advance our understanding of neurodegeneration using the multimodal neuroiamging techniques. A series of models and methods were developed and further validated through a large-scale systematic analysis on the multimodal neuroimaging datasets acquired from over 800 subjects in the Alzheimer's Disease Neuroimaging Initiative (ADNI) cohort. We designed a set of pre-processing protocols to control the quality of the datasets, then proposed a number of hand-engineered and learning-based features to model the brain morphological and functional changes associated with neurodegeneration. We further designed a multi-channel pattern analysis approach to identify the key brain regions associated with different neurodegenerative pathologies, and a cross-view pattern analysis approach to predict the synergy between these features in joint analysis of multimodal data. Finally, two clinical applications were developed to translate the research findings into improved diagnostic tools, both showing great potential in the management of Alzheimer's disease and mild cognitive impairment. A few extensions of these methods, including longitudinal neuroimaging analysis, subject-centered therapy, and brain connectome, are also demonstrated and discussed in this work.

Parts of this thesis have been published in the following journal articles:

1. W. Cai, S. Liu, L. Wen, S. Eberl, M. Fulham, D. Feng, "3D Neurological Image Retrieval with Localized Pathology-Centric CMRGLC Patterns", *The IEEE 17th International Conference on Image Processing* (**ICIP 2010**), 3201–3204 (2010). [Reproduced with Permission]

2. S. Liu, Y. Song, W. Cai, S. Pujol, R. Kikinis, X. Wang, D. Feng, "Multifold Bayesian Kernelization in Alzheimer's Diagnosis", *The 16th International Conference on Medical Image Computing and Computer Assisted Intervention* (**MICCAI 2013**), LNCS8150: 303–310 (2013). [Reproduced with Permission]

3. S. Liu, W. Cai, L. Wen, D. Feng, "Neuroimaging Biomarker based Prediction of Alzheimer's Disease Severity with Optimized Graph Construction", *IEEE International Symposium on Biomedical Imaging* (**ISBI 2013**), 1324–1327 (2013). [Reproduced with Permission]

4. S. Liu, S.Q. Liu, S. Pujol, R. Kikinis, D. Feng, W. Cai, "Propagation Graph Fusion for Multi-Modal Medical Content-Based Retrieval", *The 13th International Conference on Control, Automation, Robotics and Vision* (**ICARCV 2014**), 849–854 (2014). [Reproduced with Permission]

5. W. Cai, S. Liu, Y. Song, S. Pujol, R. Kikinis, D. Feng, "A 3D Difference-of-Gaussian-based Lesion Detector for Brain PET", *The IEEE International Symposium on Biomedical Imaging* (**ISBI 2014**), 677–680 (2014). [Reproduced with Permission]

6. S. Liu, W. Cai, L. Wen, D. Feng, S. Pujol, R. Kikinis, M. Fulham, S. Eberl, ADNI, "Multi-Channel Neurodegenerative Pattern Analysis and Its Application in Alzheimer's Disease Characterization", *Computerized Medical Imaging and Graphics* **38**, 436–444 (2014). [Reproduced with Permission]

7. S. Liu, W. Cai, S.Q. Liu, S. Pujol, R. Kikinis, D. Feng, "Subject-Centered Multi-View Feature Fusion for Neuroimaging Retrieval and Classification", *The IEEE International Conference on Image Processing* (**ICIP 2015**), 2505–2509 (2015). [Reproduced with Permission]

8. S.Q. Liu, S. Liu, F. Zhang, W. Cai, S. Pujol, R. Kikinis, D. Feng, ADNI, "Longitudinal Brain MR Retrieval with Diffeomorphic Demons Registration: What Happened to Those Patients with Similar Changes?", *The IEEE International Symposium on Biomedical Imaging* (**ISBI 2015**), 588–591 (2015). [Reproduced with Permission]

9. S.Q. Liu, N. Hadi, S. Liu, S. Pujol, R. Kikinis, D. Feng, W. Cai, "Content-based Retrieval of Brain Diffusion Magnetic Resonance Image", *The 37th European Conference on Information Retrieval Workshop on Multimodal Retrieval in the Medical Domain* (**ECIR MRMD 2015**), LNCS 9059: 54–60 (2015). [Reproduced with Permission]

10. S.Q. Liu, S. Liu, W. Cai, H. Che, S. Pujol, R. Kikinis, D. Feng, M. Fulham, ADNI, "Multi-Modal Neuroimaging Feature Learning for Multi-Class Diagnosis of Alzheimer's Disease", *IEEE Transactions on Biomedical Engineering* **62(4)**, 1132–1140 (2015). [Reproduced with Permission]

11. S. Liu, W. Cai, S.Q. Liu, F. Zhang, M. Fulham, D. Feng, S. Pujol, R. Kikinis, "Multimodal Neuroimaging Computing: A Review of the Applications in Neuropsychiatric Disorders", *Brain Informatics* **2(3)**, 167–180 (2015). [Reproduced with Permission]

12. S. Liu, W. Cai, S.Q. Liu, F. Zhang, M. Fulham, D. Feng, S. Pujol, R. Kikinis, "Multimodal Neuroimaging Computing: The Workflows, Methods and Platforms", *Brain Informatics* **2(3)**, 181–195 (2015). [Reproduced with Permission]

13. S. Liu, W. Cai, S. Pujol, R. Kikinis, D. Feng, ADNI, "Cross-View Neuroimage Pattern Analysis in Alzheimer's Disease Staging", *Frontiers in Aging Neuroscience* **8(23)**, (2016). [Reproduced with Permission]

Acknowledgements

Over the past four years, I have received support and inspiration from a great number of individuals. I would like to thank everyone who have helped me during this journey.

I would like to express my deepest appreciation to my supervisor, Assoc. Prof. Weidong Cai, for his excellent guidance and constant support. He has always encouraged me to think differently and to take advantage of my multidisciplinary background to enhance my research. He has also created many opportunities for me to meet world-renowned researchers and visit their labs. He has been a great mentor, colleague, and friend. Without his help, I would never have been able to finish this work.

I would also like to thank my associate supervisor, Prof. Dagan Feng, for providing excellent expertise and computing resources to support my Ph.D. study. Prof. Feng is the director of the Biomedical and Multimedia Information Technology (BMIT) Research Group at School of Information Technologies, University of Sydney. As one of the leading medical image analysis groups in Australia, the BMIT group comprises world-class researchers from various backgrounds with complementary skills from IT, biomedical engineering, medical imaging, and life sciences, at various stages of their careers from undergraduate students to senior professors and fellows. I am particularly indebted to Prof. Michael Fulham, Assoc. Prof. Stefan Eberl and Dr. Lingfeng Wen at Royal Prince Alfred Hospital (RPAH) for providing me strong mentorship with expertise in medical imaging research and critical clinical knowledge, together with full collaborative access to their state-of-the-art medical imaging facilities. In addition, I would like to thank my labmates Lelin Zhang, Siqi Liu, Fan Zhang, and Yang Song for brainstorming brilliant ideas in our lunchtime discussions, and Scott Lill for his great help on proofreading and polishing this work.

I am very grateful to Prof. Ron Kikinis and Dr. Sonia Pujol for offering me the opportunity to study at the Surgical Planning Lab (SPL), Harvard Medical School. During my one-year visit at SPL, I worked closely with physicians, computer scientists and engineers, and together we carried out translational research on brain white matter pathway reconstruction and post-processing. This research experience has expanded my research capability, and improved my understanding of the translational medicine research. It is my honor to work with these distinguished scientists.

I gratefully acknowledge the funding sources that made my Ph.D. work possible. My research was funded by Australia Postgraduate Award (APA), University of Sydney International Scholarship (UsydIS), Alzheimer's Australia Dementia Research Foundation (AADRF) Top-Up Scholarship, Sydney University Graduate Union North America (SUGUNA) Scholarship, National Information and Communication Technologies Australia (NICTA) Summer Scholarship, University of Sydney Postgraduate Research Travel Scheme (PRTS) Grants, and Medical Image Computing and Computer Assisted Intervention Society (MICCAI) Student Travel Grants.

Finally and most importantly, I would like to thank my family for all their love and support. I am grateful for my parents, Xuesong Zhao and Yumin Liu, who raised me and my brother, Xiangnan, and supported us in all our pursuits. I am also thankful for my parents-in-law, Chunfang Bai and Xianzhong Yuan, who helped take care of my two lovely daughters, Anne and Emilia, so that I can focus on my research. Most importantly, I would like to express my gratitude towards my beautiful wife, Shuai Yuan, who has always trusted and stood by me through good and tough times. She has made significant sacrifices for our family and deserves more honor than me. The love of my family is the greatest blessing in my life.

Author's Declaration

I declare that the work in this dissertation was carried out in accordance with the requirements of the University's Regulations and Code of Practice for Research Degree Programmes and that it has not been submitted for any other academic award. Except where indicated by specific reference in the text, the work is the candidate's own work. Work done in collaboration with, or with the assistance of, others, is indicated as such. Any views expressed in the dissertation are those of the author.

October 2015   Sidong Liu

Contents

List of Figures

List of Tables

Chapter 1
Introduction

Modern neuroimaging technologies, such as magnetic resonance imaging (MRI), positron emission tomography (PET) and electro-/magneto-encephalography (EEG/MEG), have transformed the way we study the brain [48] by providing essential anatomical and functional information about the brain in unprecedented details. Recently in the USA, President Obama launched the 'Brain Research through Advancing Innovative Neurotechnologies (BRAIN) Initiative' on his state of the union address, which refueled the interest in neuroscience research with an ambitious goal to better understand the brain over the forthcoming decade [43]. Other similar large-scale brain research projects are in progress in the European Union [8] and Asia [46].[1]

Multimodal neuroimaging is one of the most important drivers in neuroscience research, which provides essential functional and anatomical information about the brain [71]. It enables new opportunities to study the brain under normal and pathological conditions with a wide range of medical applications, such as providing clinical support to the diagnosis of neurological and psychiatric disorders, stroke, traumatic brain injuries and brain tumors [68].

This chapter aims to introduce the advances in neuroimaging technologies with a focus on their medical applications, especially in neurodegenerative disorders. Section 1.1 gives overviews of the common neuroimaging techniques, and discusses their imaging capabilities and clinical benefits. Section 1.2 describes the impact of neurodegenerative disorders and shows how they may change individuals' lives and become a global burden for our societies. Section 1.3 discusses the challenges of multimodal neuroimaging computing in neurodegeneration research. Section 1.4 summarizes the main contributions of this work towards the goal to tackle these challenges. Finally, Sect. 1.5 outlines the structure of this thesis.

---

[1]Some content of this chapter has been reproduced with permission from [68, 71].

© Springer Nature Singapore Pte Ltd. 2017
S. Liu, *Multimodal Neuroimaging Computing for the Characterization of Neurodegenerative Disorders*, Springer Theses, DOI 10.1007/978-981-10-3533-3_1

1.1   An Overview of Neuroimaging

Neuroimaging technologies vary in imaging mechanisms and capabilities. This section summarizes the characteristics of five common neuroimaging techniques, including structural MRI (sMRI), diffusion MRI (dMRI), functional MRI (fMRI), PET and EEG/MEG, and demonstrates their particular applications in the management of many brain disorders, as well as brain injuries and brain tumors.

1.1.1   Recent Advances in Neuroimaging

Neuroimaging techniques can be roughly grouped into the functional and the structural categories, depending on the information offered in the imaging data. For example, sMRI data can reveal the detailed anatomy of the brain; functional imaging, such as fMRI, PET, EEG and MEG, provides important information of the brain metabolism and neural activity. dMRI can be considered as a functional imaging technique, since it models the diffusion effects of water molecules, but it can also provide the structural information about the trajectories of brain white matter pathways. The imaging capabilities of individual neuroimaging modality can be summarized with respect to following aspects:

- SR - spatial resolution, the ability to study brain anatomy and morphology
- TR - temporal resolution, the ability to trace real-time neural activities
- SC - structural connectivity, the ability to model pathways of brain white matter
- FC - functional correlation, the ability to capture neural co-activation
- MI - molecular imaging, the ability to trace specific molecular activities
- SF - safety, the users' safety when using this neuroimaging technique
- CA - clinical availability and accessibility of this technique
- FD - future development and potential of this technique

Figure 1.1 use the polar diagrams to illustrate the capabilities of individual imaging modalities. Each axis represents an attribute, and greater length indicates better performance. Notice that indexes in this diagrams are indicative only, but not quantitative. As shown in Fig. 1.1, each technique has some unique attributes and cannot be depleted/replaced by others.

sMRI includes a range of imaging sequences - T1-weighted, T2-weighted, fluid-attenuated inversion recovery (FLAIR), proton density [99] - that provide anatomical information about the brain in details and are critical for the diagnoses of many neurological and psychiatric disorders that cause alterations to the brain structures. MRI uses various combinations of magnetic fields and radio frequencies to generate the images with no known harmful side-effects, unlike X-ray and γ-ray with ionizing radiation. In general, MRI is a safe procedure, but it might be lethal to those with implanted pacemakers and defibrillators [38]. Nevertheless, many new MRI-compatible pacemakers and defibrillators have been developed, and more are

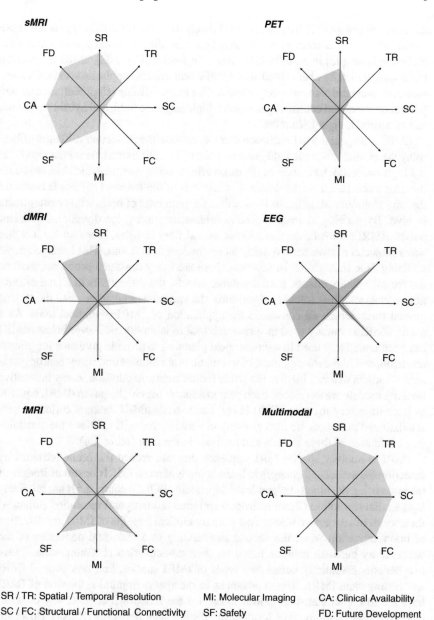

SR / TR: Spatial / Temporal Resolution   MI: Molecular Imaging   CA: Clinical Availability
SC / FC: Structural / Functional Connectivity   SF: Safety   FD: Future Development

Fig. 1.1 The strengths/limitations of sMRI (*blue*), dMRI (*green*), fMRI (*orange*), PET (*red*), EEG (*violet*) and multimodal neuroimaging (*grey*). Figure reproduced with permission from [68]

under development [22]. To date, approximately over 25,000 MRI scanners are in use worldwide [96]. The most advanced sMRI can visualize the brain at a very fine scale (0.32 mm isotropic) in an ultra-high magnetic field (9.4T) [102]. sMRI is a mature technique, which has been used in scientific and clinical applications for decades. However, there are many new opportunities for future MRI development, such as new pulse sequences and contrast agents, ultra-high magnetic field, and hybrid functional and structural imaging scanners.

dMRI is a special MRI sequence that can encode the molecular diffusion effects using the bipolar magnetic field gradient pulses [18]. Diffusion tensor imaging (DTI) is a high-order representation of diffusion effects using tensors, which can delineate the fiber tracts based on the dominate directions of the tensors [15] and is currently the only technique allowing us to visualize the pathways of brain white matter tracts in vivo. By probing at many different orientations using the directional gradient pulses, dMRI estimates the direction of axonal fiber bundles, based on the fact that water molecules move most quickly along the length of axons. This leads to longer scanning time than sMRI. In addition, there are many unsolved problems, such as the crossing-fibers effects, partial volume effects, the differences in signal estimation models and fiber tracking algorithms, the variations in datasets, and the lack of ground truth, which all constrains the application of dMRI in clinical trials. As a result, dMRI is mainly used as a research tool in laboratories. Nevertheless, dMRI has been tentatively used in neurosurgical planning to provide guidance for glioma resection, and in the management of traumatic brain injuries and many neuropsychiatric disorders that can involve the white matter tracts. In addition, many innovative imaging models are developed each year to address the challenges in dMRI, e.g., the q-space trajectory imaging (QTI) [118]. Large-scale dMRI datasets collected with standardized protocols are also growing in volume, and will facilitate the standardized evaluation of these models and methods in the near future [34].

fMRI is another special MRI sequence that can record the brain activities by detecting the associated changes in brain hemodynamics. fMRI constructs images of the brain using the blood-oxygen-level dependent (BOLD) contrast.[2] It has relatively high spatial resolution (2 mm isotropic) and medium temporal resolution (minutes) for a series of successive scans. Two primary clinical benefits of fMRI are detection of brain activation when the subject is attending to a task, and deduction of the connectivity between neurons based on their co-activation at resting state. These two benefits essentially define two types of fMRI studies, i.e., task-evoked fMRI and resting state fMRI. Recent advances in the spatiotemporal resolution of fMRI have led to higher statistical power to deduce the functional connectivity between dispersed brain regions, thus forming the resting state networks (RSNs). However, further research is required to derive and validate the biomarkers of the neurological and psychiatric disorders from the brain networks and network dynamics.

PET, the most powerful and versatile approach to study the neurotransmitter and receptor interactions, usually has lower spatiotemporal resolution than MRI and

---

[2]BLOD is closely related to cerebral blood flow (CBF), as brain function requires blood flow to supply oxygen for energy consumption by neurons.

requires the injection of radioactive tracers and exposure to ionizing radiation. Inherently, PET is a molecular imaging technique, which is exquisitely sensitive to detect the targeted molecule activities or neurotransmission processes. *2-[^{18}F]fluoro-2-deoxy-D-glucose* (FDG) is currently the most common radioactive tracer in clinical practices, which can assess the glucose metabolic levels in brain and provide support for diagnosis, staging, and evaluating treatment of cancers [121] and dementia [24, 25, 54, 57, 59, 60, 62, 64, 66, 70, 123]. A major research direction for PET is the development of specific and selective PET tracers, for example, a new tracer to detect the Tau compounds which are usually in the form of hyperphosphorylated Tau found in neurofibrillary tangles (NFTs) and cause neuron injuries and death [91, 124]. Another future direction, analogous to fMRI localization of activation, is to improve the imaging capability in dynamic assessment of neurochemical-specific brain activation [87].

EEG and MEG, which can record the synchronized activities of a population of neurons by detecting the weighted sum of their instantaneous neuronal electrical current or magnetic fluxes throughout the brain, has been heavily used in the investigations of cortical activation patterns and the event-evoked neural information pathways with ultra-high temporal resolution. Due to the simplicity and mobility of EEG/MEG systems, they have been widely used in in neurology clinics. However, EEG and MEG are constrained by their poor spatial resolution, low specificity, and inability to detect signals from the subcortical regions. Experimental studies has been carried out to integrate EEG and MEG to fMRI system [42], which offer a new imaging opportunity for the future development of EEG and MEG, i.e., EEG and MEG detect the brain activation at much finer temporal resolution than fMRI, whereas the fMRI provides the anatomical reference which compensates the inherent poor spatial resolution of EEG and MEG signals. There are a number of impediments in this thread, such as how to develop innovative approaches to operate EEG and MEG in a high magnetic field, how to better understand the correlation between BOLD signals and electrophysiological events, and how to enhance EEG/MEG signals and the fMRI signals that are acquired at the same time.

The clinical benefits of multimodal neuroimaging techniques, and the improved clinical accessibility of hybrid imaging systems, such as PET/CT [17, 112], PET/MRI [20], and PET/MRI/EEG [102], have reshaped the current neuroimaging research. Multimodal neuroimaging compensates for the inabilities of individual imaging techniques and enables a more comprehensive description of the brain. For instance, we can simultaneously analyze a task-invoked brain activation signal and its underlying information pathway using the functional data provided by fMRI and the structural information provided by dMRI tractography. With EEG/fMRI, we can improve the spatiotemporal resolution that cannot be achieved by a single imaging modality alone. Multimodal neuroimaging can also cross-validate the findings from different sources and identify their associations and patterns, e.g., causality of brain activity can be deduced by linking dynamics in different imaging readings; it can provide experimental access to determine the roles of brain structures from multiple perspectives. Lack of the computing tools is a major impediment to use of multimodal neuroimaging.

1.1.2   Applications of Neuroimaging

Neuroimaging, particularly multimodal neuroimaging, has been increasingly used in all areas of neuroscience research, including neurology, psychiatry, neurophysiology and neurosurgery. In a recent literature search on PubMed using the keywords 'multimodal AND neuroimaging' up to '31 Dec 2014', there were 1461 relevant publications retrieved from the database. Figure 1.2 shows how multimodal neuroimaging in neuroscience research expanded over the past 40 years, especially the last decade. In 2004 there were 30 related publications, and in 2014 the number of related papers jumped to more than 300 (light grey area in Fig. 1.2). Multimodal neuroimaging has many scientific and medical applications, e.g., building brain computer interfaces [80], detecting neural activities [65], tracing information pathways [73], mapping brain functions to structures [19, 86, 90], monitoring pharmacological treatments [77, 119], evaluating image-guided therapies [82, 98, 108], and etc.

One of the important clinical application of multimodal neuroimaging lies in the provision of essential functional and anatomical information about the brain for the management of neurological and psychiatric disorders [23, 43]. This was demonstrated by our subsequent search on these 1461 PubMed publications, using the following keywords: *'(multimodal **AND** neuroimaging) **AND** (neuropsychiatric **OR** neurological **OR** psychiatric)'*. A substantial proportion of the retrieved publications are targeting on the neuropsychiatric disorders, and their number dramatically increased from 10 to 121 over the last decade (dark grey area in Fig. 1.2).

These multimodal neuroimaging approaches used in these neuropsychiatric studies can be classified into three categories based on the applied neuroimaging techniques, i.e., the structural-structural imaging combinations, the functional-functional imaging combinations, and the structural-functional imaging combinations. Each

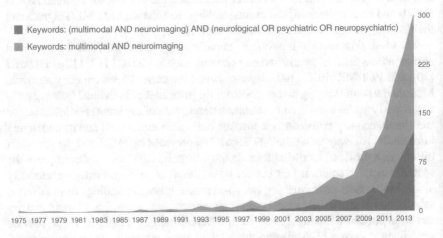

Fig. 1.2 The explosive growth of use and development of multimodal neuroimaging over the past 40 years (1975–2014). Figure reproduced with permission from [68]

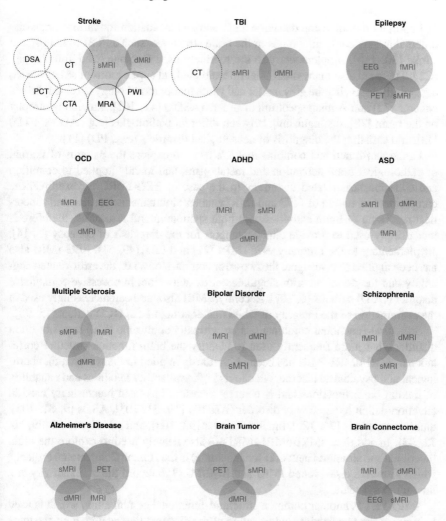

Fig. 1.3 Medical applications of multimodal neuroimaging in neuropsychiatric disorders. Figure reproduced with permission from [68]

combination has particular strengths and applications in these neuropsychiatric disorders, as shown in Fig. 1.3. Colors of the circles indicate individual neuroimaging techniques, as same as in Fig. 1.1. Sizes of the circles indicate the prevalence of these techniques in specific neuropsychiatric disorders. Note the sizes are merely indicative but not quantitative.

Structural-structural combinations, in brief, are commonly used to extract the morphological features from the multimodal data. The sMRI-dMRI methods dominate the structural-structural category, as they combines the microstructural and macrostructural information in both brain grey matter and white matter. It has become an important tool in lesion detection and treatment evaluation for various brain disorders that may cause alterations in the brain structures, such as TBI [44, 114] and stroke [30, 110], supplementary to the routinely used Computerized Tomography (CT). There are also many other scientific and medical applications of sMRI-dMRI, e.g., jointly analyzing the grey matter and white matter morphology in schizophrenia patients [72] and Autism spectrum disorders (ASDs) [41, 101], simulating atrophy on the brain [78], distinguishing between different pathologies (e.g., AD vs. FTD) [13], and assisting the diagnosis of neurological disorders (e.g., PD) [11].

Functional-functional combinations are used to explore the patterns of normal or pathological brain activation and metabolism, and mainly applied to cognition and consciousness-related disorders. In the case of EEG-fMRI, for example, the complementary nature of EEG and fMRI enables simultaneous cortical and subcortical recording of brain activities with high spatiotemporal resolution, therefore is increasingly used to provide clinical support for the diagnosis of epilepsy [1, 14], supplementary to the routinely used sMRI [21] and PET [49, 51]. EEG-fMRI also has been applied to investigate the two error commissions, i.e., the error-related negativity and the dorsal anterior singulate cortex activation, in obsessive-compulsive disorder (OCD) patients [6, 107]. Recently, dMRI also has been increasingly used in the examination of the integrity of white matter structures in OCD patients [5, 95].

Structural-functional combinations are virtually applicable to all forms of brain disorders, but more frequently used to identify the brain function-structure correlations. sMRI-dMRI-fMRI has been ubiquitously applied in the research on neurological and psychiatric disorders due to its high availability in clinics and capability of linking brain functions and structures, therefore has been increasingly used in attention-deficit hyperactivity disorder (ADHD) [10, 31, 103], ASDs [9, 81, 105], multiple Sclerosis [74, 109], bipolar disorder [92, 106], and schizophrenia [29, 50, 72, 93]. In addition, sMRI-dMRI-fMRI are also heavily used to explore the brain functional and structural networks (connectome) in the Human Connectome Project,[3] which also uses task-evoked fMRI and EEG/MEG to record the brain activity at a finer temporal scale.

sMRI-PET is another common structural-functional combination, which is used to improve the localization and targeting of the diseased tissues with high accuracy and sensitivity. In the case of Alzheimer's disease (AD), for example, sMRI and PET, usually using the prevalent ^{18}F-FDG or amyloid tracers, are used to measure the structural changes (e.g., atrophy) and functional changes (e.g., hypometabolism, amyloid plaques and NFTs) [26, 56, 61, 67, 122] in the brain. Recently, dMRI [88, 94], fMRI [42], or both [45, 117, 125], have also been increasingly used in the evaluation of AD. sMRI-PET also has been used to locate and mark the lesional areas, e.g., tumor and edema, in neuro-oncology cases. When sMRI is combined with PET, with the standard ^{18}F-FDG, it shows a great potential for more precise delineation of the tumor boundary than using sMRI alone [20, 85]. In brain tumor surgery, dMRI can be combined with sMRI and PET for preoperative surgical planning and intraoperative surgical navigation. For instance, in a recent brain tumor study, dMRI

---

[3] www.neuroscienceblueprint.nih.gov/connectome.

was used to predict tumor infiltration in patients with gliomas [33]. Another example was given by Tempany et al. [108], where dMRI tractography is jointly displayed with sMRI tumor to give a complete brain map for surgical planning.

1.2 Neurodegenerative Disorders

1.2.1 A Disabling Condition to Patients

Neurodegenerative disorders, such as AD and PD, are believed to be caused by the death of brain cells, and can have a profound impact on the lives of patients. As their cognitive abilities in memory, reasoning, planning, speaking, and behavior decline, people with neurodegenerative disorders gradually lose control of many essential features of their lives, and these conditions eventually lead to disability and death [47]. Neurodegenerative disorders, together with other neurological, mental and behavioral disorders, represent the most disabling category, based on a systematic analysis of descriptive epidemiology of 291 diseases and injuries from 1990 to 2010 for 187 countries [83]. As shown in Fig. 1.4, neuropsychiatric disorders caused the largest number of years lost due to illness, disability and early death measured by the disability-adjusted life years (DALYs).[4] As of 2010 in US, the average DALYs percentage related to neuropsychiatric disorders was 18.7%, which has overtaken cardiovascular and circulatory diseases (16.8%), neoplasms (15.1%), musculoskeletal disorders (11.8%), and other non-communicable diseases. Neurodegenerative disorders are associated with a large increase (50%) in DALYs. Such increase suggests an important trend that neurodegenerative disorders are contributing to the growing challenge of chronic disability, and the same trend is also evidenced in other Organization for Economic Co-operation and Development (OECD) countries. According to the statistics released by Australian Bureau of Statistics (ABS), AD and dementia-related disorders has become the third leading cause of death in Australia as of 2011. More importantly, the number of deaths due to this cause has increased by 126% during the period from 2002 to 2011 [3].

Neurodegenerative disorders also have a profound impact to the patients' carers. Around 94% of the people with these conditions need help with at least one basic everyday task, e.g., 24% of patients living in households need help to express their feelings and emotions, 35% need help to interact with others, 58% need help with reading and writing, and 59% need help with decision-making and problem solving [2]. Patients living in health establishments, such as hospitals, nursing homes, aged care hostels, or retirement villages, tend to have more advanced dementia than their counterparts living in households. They require constant care and supervision from

---

[4]DALYs is a summary metric of population health, which is the sum of years of life lost due to premature mortality and years lived with disability. DALYs represents a health gap, and as such, measures the state of a population's health compared to a normative goal that is for individuals to live the standard life expectancy in full health [83].

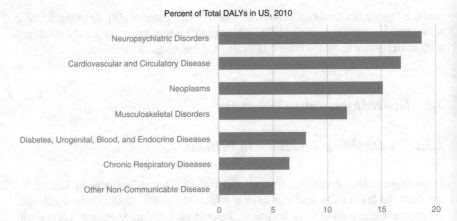

Fig. 1.4 The disability-adjusted life-years (DALYs) of 291 diseases and injuries based on the systematic analysis of descriptive epidemiology from 1990 to 2010 in US. Figure reproduced with permission from [68]

their caregivers, and this demands the caregiver to be heavily involved, be aware of the symptoms, treatments and progression of the disease, and be ready to various situations, e.g., patients may have dramatical emotion changes, becoming very demanding or depressed.

1.2.2 An Economic Burden to the Society

Neurodegenerative disorders will also be the most costly global burden over the coming decades, and the socioeconomic burden of the disorders will worsen as longevity increases. In the case of AD alone, in 2006, there were 26.6 million patients diagnosed with AD worldwide, including 56% of the cases in the early stage of the disease. This number is predicted to grow fourfold to more than 100 million in 2050 [23]. The current spending on AD is about 157 - 215 billion US dollars per year in US alone, and the projected spending in 2050 will be over one trillion US dollars per year in US, as shown by the dark grey bars in Fig. 1.5 [7].

Research has indicated that dietary and lifestyle changes, for instance, quitting smoking and doing more exercise, may lower the risk of developing neurodegenerative disorders. However, there are no cures for these disorders. Current medical interventions may only slow or halt its progression in the early stage of the disease, but cannot reverse the onset of dementia once the symptoms appear. As predicted by the Alzheimer's Association, should the onset of dementia delayed for 5 years, the estimated cost to treat AD patients will drop to 600 billion per year in 2050 in US [7], as shown by the light grey bars in Fig. 1.5. Therefore, the early diagnosis of neurological disorders is vitally important for patients at higher risk of develop-

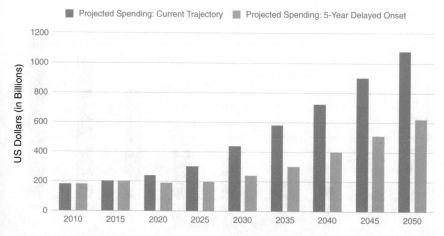

Fig. 1.5 The projected annual cost of Alzheimer's disease over the period between 2015 and 2050 in US [43]

ing dementia to receive proper treatments as early in the course as possible, prior to irreversible symptoms appear. Ongoing research into causes and treatments may eventually lead to the medical interventions that could cure patients in the future, but there is still a long way to go.

1.3 Main Challenges

Neuroimaging has been increasingly used in the diagnosis of neurodegenerative disorders. However, great challenges exist due to the complexity of disease pathologies, since neurodegenerative disorders are generally progressive and appear in many different forms. From the neuroimaging side, the analysis and processing of neuroimaging data, especially multimodal neuroimaging data, is a very challenging task, requiring sophisticated processing, such as artifact correction, image registration and segmentation, feature extraction, pattern recognition and visualization, as well as the domain knowledge to understand the biophysical and biochemical information in the data.

1.3.1 Complexity of Disease Pathologies

Neurodegenerative disorders are highly age-related, but it is not part of the normal aging process. Although there are young people identified as having these conditions at their late 20s, the onset of dementia is usually in old age, 96% are 65 years or older. In Australia, the proportion identified as having dementia increase steadily with age

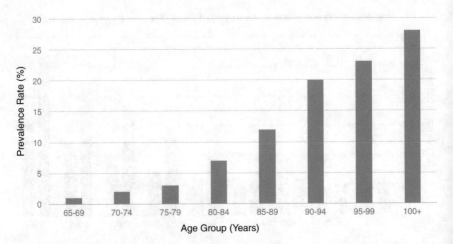

Fig. 1.6 The prevalence rate of dementia in different age groups in Australia, 2009 [4]

from 1% in the 65 - 69 group to 28% of the centenarians group [4], as shown in Fig. 1.6. A cognitively normal people over 75 years old may have an average low metabolic activity compared to a 30 years' old patient with dementia, therefore, wrong assertions can be made if the influence of age is ignored. Similarly, other factors, such as sex, education and genotypes, need to be taken into consideration when designing the neurodegeneration related studies.

There are many forms of neurodegenerative disorders. Alzheimer's disease (AD) is the most common neurodegenerative disorder among aging people, which accounts for nearly 70% of all dementia cases [47]. Other common neurodegenerative disorders include frontotemporal dementia (FTD), vascular dementia (VD), Parkinson's disease (PD), and dementia with lewy bodies (DLBD). There are also neurological disorders seen at a low frequency, such as motor neurone disease (MND), Jacob-Creutzfeldt disease (CJD), corticobasal degeneration (CBD), progressive supranuclear plasy (PSP), and multiple system atrophy (MSA).

Neurodegenerative disorders are also progressive. In the early stage, the patients may feel having less energy and spontaneity, then experiencing noticeable decline in memory, language and other cognitive abilities, and becoming depressed. Patients with these signs of neurodegeneration are usually diagnosed as the mild cognitive impairment (MCI), which does not notably interfere with daily activities, but leads to a higher risk of developing dementia [35, 97]. A recent ADNI study suggests that MCI is indeed a heterogeneous group, which includes a substantial proportion of non-AD pathologies, such as VD and FTD [84]. As symptoms of neurodegeneration gradually deteriorate over years, the patients gradually lose ability to perform everyday tasks, such as expressing their feelings and emotions, interacting with others, reading and writing, decision-making and problem solving. Finally, they may lose the ability to perform basic functions, such as chewing and swallowing, and stay in bed until death.

1.3.2 Difficulties in Neuroimaging Computing

The use of neuroimaging data and development of neuroimaging technologies have spurred a parallel development of brain image computing, which focuses on the computational processing and analysis of neuroimaging data. Neuroimaging computing is inherently an interdisciplinary research area at the intersection of neuroscience, computer science, mathematics and statistics. The main goal of neuroimaging computing is to extract relevant information pertaining to the neuroimaging data, and apply such information to clinical care or neuroscience research. However, neuroimaging computing is a very challenging task, requiring sophisticated design and domain knowledge to analyze the neuroimaging data in different modalities. This section provides an overview of the neuroimaging computing methods and popular packages. More details can be found in Chap. 2, Sect. 2.1.2.

The sMRI computing workflows usually involve artifact correction, skull stripping, tissue segmentation, surface reconstruction [37], followed by brain morphometry analysis, such as voxel-/tensor- based morphometry (VBM/TBM) [12]. If the brain is segmented into different regions of interest (ROIs) based on a brain template, then the analysis can be carried out at the ROI level. Popular brain templates include the International Consortium for Brain Mapping (ICBM) template [76] and the Automated Anatomical Labeling (AAL) template [115] in the Montreal Neurological Institute (MNI) coordinates [36].

The computing of dMRI starts with the estimation of the principle fiber directions in each voxel, usually represented by a tensor. Advanced fiber orientation estimation methods include the ball-and-stick model [16], the constrained spherical deconvolution (CSD) model [111], q-ball imaging (QBI) [113], diffusion spectral imaging (DSI) [116], generalized q-sampling imaging (GQI) [120], and QBI with Funk-Radon and Cosine Transform (FRACT) [40]. Various voxel-wise coefficients, such as mean diffusivity (MD), radial diffusivity (RD), axial diffusivity (AXD) and fractional anisotropy (FA) [75], can be extracted [100]. Fiber tracking algorithms [79] are then applied to reconstruct the major trajectories of the brain white matter pathways, i.e., the tractography [32]. Tractography enables the analysis of brain white matter morphology and structural connectome [89] (Fig. 1.7).

In fMRI computing, body movement is a serious issue, thereby requires serial alignment, spatial normalization and smoothing for head-motion correction and slice timing correction, followed by activation parameter estimation. Resting state fMRI can be used to deduce the co-activation patterns of dispersed brain areas, thus enabling the analysis of functional connectome. The primary use of task-evoked fMRI is to reveal the correlation between the brain activation patterns and the cognitive functions, such as perception, language, memory and emotion.

PET computing also requires serial alignment, spatial normalization and smoothing for body movement correction, and then estimates the parameters pertaining to the radioactive tracers. The common parameters include standard uptake value ratios (SUVR) [28, 52], cerebral metabolic rate of glucose consumption (CMRGlc) [24, 104] and hypo-metabolic convergence index (HCI) [27] for FDG-PET, and amyloid convergence index (ACI) [27] for amyloid PET.

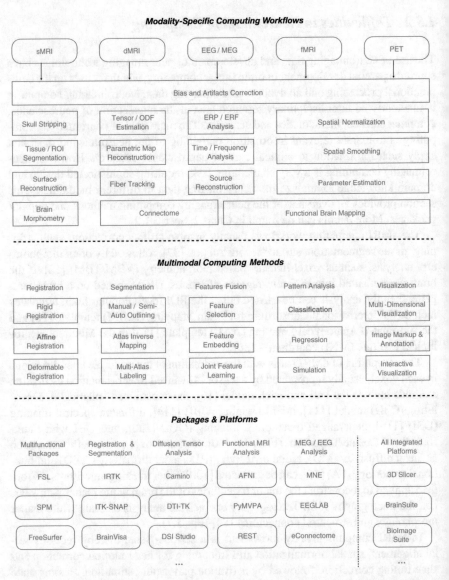

Fig. 1.7 The modality-specific and multimodal neuroimaging computing workflows, and the neuroimaging computing packages and platforms. Figure reproduced with permission from [71]

The practice of EEG/MEG usually contains the following steps. The artifacts or unwanted data components that contaminate the signals are usually manually or automatically removed, followed by the computation of the event-related potentials (ERP) or event-related fields (ERF), which were transformed into the time-frequency domain. The signals are then measured at the sensor level, and subsequently used to infer the neural sources, which enables the investigation of the information flow and interaction between brain areas [39].

Multimodal neuroimaging analysis is a more challenging task due to large inter-modality variances in spatiotemporal resolution and merge the biophysical/bio-chemical information in images. In addition to the modality-specific computing, multimodal neuroimaging computing requires more sophisticated computing in pre-processing, image registration, segmentation, feature extraction, pattern recognition and visualization.

1.4 Main Contributions

This study aims to tackle the challenges mentioned in previous section by extracting the clinically relevant information from the multimodal neuroimaging data, exploring the patterns of the neurodegenerative pathologies, deducing the disorder biomarkers, and developing clinical applications for transforming the imaging-based findings into improved diagnostic tools. A series of models and methods were proposed in this study and the main contributions of this study can be summarized using a four-layer architecture, each with a specific goal.

Data Computing The data computing layer focuses on neuroimaging data acquisition and pre-processing. In this study, we recruited four groups of subjects from the public ADNI database to ensure sufficient coverage of neurodegeneration progression stages, subject samples, and imaging modalities. These datasets includes the MRI subset (758 subjects), the PET subset (352 subjects), the dMRI subset (233 subjects), and the multimodal MRI-PET subset (331 subjects). For each subset, we further designed a set of pre-processing protocols to ensure the quality of the datasets, including artifact correction, spatial and functional normalization, image registration, brain segmentation and parcellation, and parameter estimation. The details of the subject selection criteria and image pre-processing protocols for each subset can be found in Chap. 3.

Feature Representation The feature representation layer focuses on neuroimaging-based features that can best depict the disease pathologies, such as atrophy and hypo-metabolism. Two unique brain morphological features, i.e., Convexity Ratio (CNV) and Solidity Ratio (SLD), were proposed to depict the brain atrophy in MRI. Both CNV and SLD showed complementary characteristics to the conventional brain morphological features, i.e., grey matter volume (GMV) and local cortical gyrification index (LGI), in the analysis brain morphometry. In addition, a novel 3D Difference-of-Gaussian (DoG) lesion detector with three lesion-centric feature descriptors (DoG-Mean, DoG-Contrast, DoG-Z-Score) was designed for PET to detect the hypo-metabolism in the brain and enable the quantitative analysis of the brain neurodegeneration with a focus on the lesion areas, rather than the entire brain regions as with the mean index (M-IDX) and fuzzy index (F-IDX). To take the advantage of multimodal data, we further proposed a novel feature learning framework with the deep learning architecture. The deep-learning-based features demonstrated higher statistical power than the hand-engineered features in extracting the complex correlations between brain ROIs, features, and modalities, and was capable of detecting the subtle differences between MCI patients who converted to AD (converters) and those who did not convert (non-converters). The details of the feature modeling and learning methods can be found in Chap. 4.

Pattern Analysis The pattern analysis layer focuses on patterns of disease pathologies on whole brain. The goal of pattern analysis is to identify the most discriminant brain regions associated with the disease, and to obtain a comprehensive picture of the disease pathology. A multi-channel pattern analysis method was designed to identify the disease-specific regions by fusing the pathological patterns derived with different analysis methods, i.e., Elastic Net, Support Vector Machine, and Two-One-Sided T-Test, and then rating the regions using a vote-rule fusion method. The proposed multi-channel method was capable of compensating the deficiency of individual analysis method, and effectively enhanced the neurodegeneration pattern-based retrieval. In addition, a cross-view pattern analysis approach was developed to investigate the correlation between the pathological patterns obtained from different features, i.e., CNV, SLD, GMV, LGI, DoG-Mean, DoG-Contrast, DoG-Z-Score, M-IDX and F-IDX. The proposed cross-view method made use of the mutual divergence to quantize the variance of these features, and showed great potential in predicting the synergy of the multimodal features in disease characterization. Chapter 5 provides the technical details of pattern analysis methods and analyzes their derived patterns.

Application Development The application development layer focuses on the translational applications that transform the discoveries into diagnostic tools. A knowledge-encoded graph-cut algorithm was developed to integrate the domain knowledge in disease staging with a new global-cost to encode the domain knowledge and a redesigned the objective function. This optimized graph-cut algorithm significantly improved the distinction between patient groups by minimizing the intra-group variance and maximizing the inter-group variance. In addition, a novel multifold Bayesian kernelization (MBK) algorithm was developed for multimodal classification of AD, MCI and normal controls. MBK is a novel classifier-independent model which infers the patient's diagnosis by synthesizing the output prediction probabilities of multimodal biomarkers. One prominent advantage of MBK is that the prediction can be made based on arbitrary number of biomarkers without a need of re-training the model, and the training outputs, i.e., weights of the biomarkers, are transferable to other classifiers as references for feature selection. The methods and experimental results of the optimized graph-cut algorithm and the MBK algorithm can be found in Chap. 6. Furthermore, a propagation graph fusion (PGF) algorithm was developed to provide clinical decision support through subject-centered content-based neuroimaging retrieval. PGF is an unsupervised method, which requires no prior knowledge of the features or the query subjects, and is able to adaptively reshape the connections between the subjects according to query, thereby find more relevant subjects. We further enhanced the PGF algorithm by enforcing the conformity of the retrieval results across multiple views using the geometric-mean based fusion method. The improved PGF showed great potential in multimodal data management. Chapter 7 introduces the basic and the improved PGF methods and demonstrates the use of them in content-based retrieval of the multimodal neuroimaging data in ADNI database.

1.5 Structure of Thesis

This work presents the models and methods as highlighted in Sect. 1.4, and also demonstrates their effectiveness in the characterization of neurodegenerative disorders through a systematic analysis on the ADNI database. The rest of this thesis is structured as follows.

Chapter 2 reviews the previous works that related to this study. These studies are also summarized in the four layers, i.e., data computing (Sect. 2.1), feature representation (Sect. 2.2), pattern analysis (Sect. 2.3), and application development (Sect. 2.4). Some content of this chapter has been reproduced with permission from [24, 55].

Chapter 3 introduces the databases and pre-processing protocols used in this study, including the ADNI MRI subset (Sect. 3.1), PET subset (Sect. 3.2), dMRI subset (Sect. 3.3), and MRI-PET subset (Sect. 3.4). Some content of this chapter has been reproduced with permission from [53, 58, 61, 63].

Chapter 4 focuses on the descriptions of feature extraction and evaluation methods. The proposed feature descriptor to encode the brain morphological information is described in Sect. 4.1. The method of extracting the lesion-centric functional features is described in Sect. 4.2. The deep-learning framework for multimodal feature extraction is presented in Sect. 4.3. Some content of this chapter has been reproduced with permission from [25, 56, 63, 65].

Chapter 5 elaborates on the pattern analysis methods in two streams, based on the analysis methods or the feature descriptors. Section 5.1 provides the details of the channel-based pattern analysis methods and further discusses their findings. Section 5.2 presents the view-based analysis methods and results. Some content of this chapter has been reproduced with permission from [63, 66].

Chapter 6 demonstrates the clinical applications for improving the staging and prediction of AD. A domain knowledge-encoded graph-cut algorithm is introduced in Sect. 6.1. Section 6.2 presents the MBK algorithm, a novel classifier-independent AD prediction model which infers the patient's diagnosis by synthesizing the output diagnosis probabilities of multimodal biomarkers. Some content of this chapter has been reproduced with permission from [58, 67].

Chapter 7 introduces another clinical application for providing clinical decision support by improving the content-based retrieval of multimodal neuroimaging data. Section 7.1 introduces the basic PGF method and summarizes the retrieval results based on the ADNI MRI-PET subset. Section 7.2 presents an improved PGF method and demonstrates the improvements over original PGF on the ADNI MRI and dMRI subsets. Some content of this chapter has been reproduced with permission from [53, 61, 69].

Chapter 8 concludes the research findings in neurodegeneration based on the analyses in Chaps. 3 to 7 and further outlines the future directions of multimodal neuroimaging research. Some content of this chapter has been reproduced with permission from [68, 71].

Two appendixes are also provided to help the readers quickly look up the specialized vocabulary. Appendix A lists all abbreviations and acronyms appeared in this work, and Appendix B gives the detailed definitions of the brain regions in the ICBM template.

References

1. Abela, E., et al. (2014). Neuroimaging of epilepsy: Lesions, networks, oscillations. *Clinical Neuroradiology, 24,* 5–15.
2. ABS. (2012). Australia Social Trends cat.no.4102.0 (Australian Bureau of Statistics, Canberra).
3. ABS. (2011). Causes of Death, Australia cat. no. 3303.0 (Australian Bureau of Statistics, Canberra).
4. ABS. (2009). Disability, Ageing and Carers, Australia: Summary of Findings cat. no. 4430.0 (Australian Bureau of Statistics, Canberra).
5. Agam, Y., Greenberg, J. L., Isom, M., Falkenstein, M. J., et al. (2014). Aberrant error processing in relation to symptom severity in obsessive-compulsive disorder: A multimodal neuroimaging study. *NeuroImage: Clinical, 5,* 141–151.
6. Agam, Y., Vangel, M., Roffman, J. L., Gallagher, P. J., et al. (2014). Dissociable genetic contributions to error processing: A multimodal neuroimaging study. *PLoS ONE, 9,* e101784.
7. Alzheimer's Association. (2015). Changing the Trajectory of Alzheimer's Disease: How a Treatment by 2025 Saves Lives and Dollars.
8. Amunts, K., Linder, A., & Zilles, K. (2014). The human brain project: Neuroscience perspectives and German contributions. *e-Neuroforum, 5,* 43–50.
9. Anagnostou, E., & Taylor, M. J. (2011). Review of neuroimaging in autism spectrum disorders: What have we learnt and where we go from here. *Moelcular Autism, 2,* 1–9.
10. Anderson, A., Douglas, P. K., Kerr, W. T., Haynes, V. S., et al. (2014). Non-negative matrix factorization of multimodal MRI, fMRI and phenotypic data reveals differential changes in default mode subnetworks in ADHD. *NeuroImage, 102,* 207–219.
11. Aquino, D., Contarino, V., Albanese, A., et al. (2013). Substantia nigra in Parkinson's disease: A multimodal MRI comparison between early and advanced stages of the disease. *Neurological Sciences, 35,* 753–758.
12. Ashburner, J., & Friston, J. K. (2000). Voxel-based morphometry - The methods. *NeuroImage, 11,* 805–821.
13. Avants, B., Cook, P., Ungar, L., Gee, J., & Grossman, M. (2010). Dementia induces correlated reduction in white matter integraty and cortical thickness: A multivariate neuroimaging study with sparse canonical correlation analysis. *NeuroImage, 50,* 1004–1016.
14. Bagshaw, A. P., Rollings, D. T., Khalsa, S., & Cavanna, A. E. (2014). Multimodal neuroimaging investigations of alternations to consciousness: The relationship between absence epilepsy and sleep. *Epilepsy & Behavior, 30,* 33–37.
15. Basser, P., Mattiello, J., & LeBihan, D. (1994). MR diffusion tensor spectroscopy and imaging. *Biophysical Journal, 66,* 259–267.
16. Behrens, T., Woolrich, M., Jenkinson, M., Johansen-Berg, H., et al. (2003). Characterization and propagation of uncertainty in diffusion-weighted MR imaging. *Magnetic Resonance in Medicine, 50,* 1077–1088.

17. Beyer, T., et al. (2000). A combined PET/CT scanner for clinical oncology. *Journal of Nuclear Medicine, 41,* 1369–1379.
18. Bihan, D., Mangin, J., Poupon, C., Clark, C., et al. (2001). Diffusion tensor imaging: Concepts and applications. *Journal of Magnetic Resonance Imaging, 13,* 534–546.
19. Binder, J. R., Desai, R. H., Graves, W. W., & Conant, L. L. (2009). Where is the semantic system? A critical review and meta-analysis of 120 functional neuroimaging studies. *Cerebral Cortex, 19,* 2767–2796.
20. Bisdas, S., et al. (2010). Switching on the lights for real-time multimodality tumor neuroimaging: The integrated positron-emission tomography/MR imaging system. *American Journal of Neuroradiology (AJNR), 31,* 610–614.
21. Bonilha, L., & Keller, S. S. (2015). Quantitative MRI in refractory temporal lobe epilepsy: Relationship with surgical outcomes. *Quantitative Imaging in Medicine and Surgery, 5,* 204–224.
22. Bovenschulte, H., et al. (2012). MRI in patients with pacemakers - Overview and procedural management. *Deutsches Arzteblatt International, 109,* 270–275.
23. Brookmeyer, B., Johnson, E., Ziegler-Graham, K., & Arrighi, H. (2007). Forecasting the global burden of Alzheimer's disease. *Alzheimer's & Dementia, 3,* 186–191.
24. Cai, W., et al. (2010). 3D Neurological image retrieval with localized pathology-centric CMR-Glc patterns. In *The 17th IEEE international conference on image processing (ICIP)* (pp. 3201–3204). IEEE.
25. Cai, W., et al. (2014). A 3D difference of Gaussian based lesion detector for brain PET. In *IEEE international symposium on biomedical imaging: From nano to macro (ISBI)* (pp. 677–680). IEEE.
26. Che, H., et al. (2014). Co-neighbor multi-view spectral embedding for medical contentbased retrieval. In *IEEE international symposium on biomedical imaging: From nano to macro (ISBI)* (pp. 911–914). IEEE.
27. Chen, K., Ayutyanont, N., Langbaum, J. B., Fleisher, A. S., Reschke, C., et al. (2011). Characterizing Alzheimer's disease using a hypometabolic convergence index. *NeuroImage, 56,* 52–60. ISSN: 1053-8119.
28. Clark, C. M., et al. (2012). Cerebral PET with florbetapir compared with neuropathology at autopsy for detection of neuritic amyloid-OE $\leq$ plaques: A prospective cohort study. *The Lancet Neurology, 11,* 669–678. ISSN: 1474-4422.
29. Cooper, D., Barker, V., Radua, J., Fusar-Poli, P., & Lawrie, S. M. (2014). Multimodal voxel-based meta-analysis of structural and functional magnetic resonance imaging studies in those at elevated genetic risk of developing schizophrenia. *Psychiatry Research: Neuroimaging, 221,* 69–77.
30. Copen, W. A. (2015). Multimodal imaging in acute ischemic stroke. *Current Treament Options in Cardiovascular Medicine, 17,* 1–17.
31. Dai, D., Wang, J., Hua, J., & He, H. (2012). Classification of ADHD children through multimodal magnetic resonance imaging. *Frontiers in Systems Neuroscience, 6,* 1–8.
32. Durrieman, S., Pennec, X., Trouve, A., & Ayache, N. (2009). Statistical models of sets of curves and surfaces based on currents. *Medical Image Analysis, 13,* 793–808.
33. Durst, C. R., Raghavan, P., Shaffrey, M. E., Schiff, D., et al. (2014). Multimodal MR imaging model to predict tumor infiltration in patients with gliomas. *Neuroradiology, 56,* 107–115.
34. Essen, D. C. V., Smith, S. M., Barch, D. M., Behrens, T. E., Yacoub, E., et al. (2013). The WU-Minn human connectome project: An overview. *NeuroImage, 80,* 62–79.
35. Fan, Y. (2011). In T. Liu, D. Shen, L. Ibanez, & X. Tao (Eds.), *Ordinal ranking for detecting mild cognitive impairment and Alzheimer's disease based on multimodal neuroimages and CSF biomarkers* (Vol. 7012, pp. 44–51). Multimodal brain image analysis (MBIA) Berlin, Heidelberg: Springer. ISBN: 978-3-642-24445-2.
36. Fonov, V., Evans, A., Botteron, K., Almli, C., et al. (2010). Unbiased average age-approapriate atlases for pediatric studies. *NeuroImage, 54,* 313–327.
37. Geffroy, D., et al. (2011). BrainVISA: A complete software platform for neuroimaging. In *Python in neuroscience workshop.*

38. Gotte, M., Russel, I., de Roest, G., Germans, T., Veldkamp, R., et al. (2010). Magnetic reso-
     nance imaging, pacemakers and implantable cardioverter-defibrillators: Current situation and
     clinical perspective. *Netherlands Heart Journal, 18,* 31–37.

39. Gross, J., Baillet, S., Barnes, G. R., Henson, R. N., Hillebrand, A., et al. (2013). Good practice
     for conducting and reporting MEG research. *NeuroImage, 65,* 349–363.

40. Haldar, J. P., & Leahy, R. M. (2013). Linear transforms for fourier data on the sphere: Appli-
     cation to high angular resolution diffusion MRI of the brain. *NeuroImage, 71,* 233–247.

41. Hasan, K. M., Walimuni, I. S., & Frye, R. E. (2013). Global cerebral and regional multimodal
     neuroimaging markers of the neurobiology of autism. *Journal of Child Neurology, 28,* 874–
     885.

42. He, X., Qin, W., Liu, Y., Zhang, X., et al. (2013). Abnormal salience network in normal aging
     and in amnestic mild cognitive imapirment and Alzheimer's disease. *Human Brain Mapping,
     35,* 3446–3464.

43. Insel, T. R., Landis, S. C., & Collins, F. S. (2013). The NIH BRAIN initiative. *Science, 340,*
     687–688.

44. Irimia, A., Chambers, M. C., Alger, J. R., Filippou, M., Prastawa, M. W., et al. (2011).
     Comparison of acute and chronic traumatic brain injury using semi-automatic multimodal
     segmentation of MR volumes. *Journal of Neurotrauma, 28,* 2287–2306.

45. Jacobs, H. I., Gronenschild, E. H., Evers, E. A., et al. (2015). Visuospatial processing in early
     Alzheimer's disease: A multimodal neuroimaging study. *Cortex, 64,* 394–406.

46. Jiang, T. (2013). Brainnetome: A new-ome to understand the brain and its disorders. *Neu-
     roImage, 80,* 263–272.

47. Kalaria, R. N., et al. (2008). Alzheimer's disease and vascular dementia in developing coun-
     tries: Prevalence, management, and risk factors. *The Lancet Neurology, 7,* 812–826. ISSN:
     1474-4422.

48. Kikinis, R., Pieper, S. D., & Vosburgh, K. (2014). 3D slicer: A platform for subject-specific
     image analysis, visualization, and clinical support. In F. A. Jolesz (Ed.), *Intraoperative imag-
     ing and image-guided therapy 3* (pp. 277–289). New York: Springer.

49. Knopman, A. A., Wong, C. H., Stevenson, R. J., et al. (2015). The relationship between neu-
     ropsychological functioning and FDG-PET hypometabolism in intractable mesial temporal
     lobe epilepsy. *Epilepsy & Behavior, 44,* 136–142.

50. Kochunov, P., Chiappelli, J., Wright, S. N., Rowland, L. M., et al. (2014). Multimodal white
     matter imaging to investigate reduced fractional anisotropy and its age- related decline in
     schizophrenia. *Psychiatry Research: Neuroimaging, 223,* 148–156.

51. La Fougere, C., Rominger, A., Forster, S., Geisler, J., & Bartenstein, P. (2009). PET and
     SPECT in epilepsy: A critical review. *Epilepsy & Behavior, 15,* 50–55.

52. Landau, S. M., et al. (2013). Comparing positron emission tomography imaging and cere-
     brospinal fluid measurements of beta-amyloid. *Annals of Neurology, 74,* 826–836. ISSN:
     1531-8249.

53. Liu, S. Q., et al. (2015). Content-based retrieval of brain diffusion magnetic resonance image.
     Multimodal retrieval in the medical domain (Vol. 9059). Switzerland: Springer.

54. Liu, S. Q., et al. (2014). High-level feature based PET image retrieval with deep learning
     architecture. *Journal of Nuclear Medicine, 55,* 2018.

55. Liu, S. Q., et al. (2015). Longitudinal brain MR retrieval with diffeomorphic demons registra-
     tion: What happened to those patients with similar changes?. In *IEEE international symposium
     on biomedical imaging: From nano to macro (ISBI)* (pp. 588–591). IEEE.

56. Liu, S. Q., et al. (2015). Multi-modal neuroimaging feature learning for multi-class diagnosis
     of Alzheimer's disease. *IEEE Transactions on Biomedical Engineering, 62,* 1132–1140.

57. Liu, S., Cai, W., Wen, L. & Feng, D. (2012). Multiscale and multiorientation feature extraction
     with degenerative patterns for 3D neuroimaging retrieval. In *The 19th IEEE international
     conference on image processing (ICIP)* (pp. 1249–1252). IEEE.

58. Liu, S., Cai, W., Wen, L. & Feng, D. (2013). Neuroimaging biomarker based prediction
     of Alzheimer's disease severity with optimized graph construction. In *IEEE international
     symposium on biomedical imaging: From nano to macro (ISBI)* (pp. 1324–1327). IEEE.

59. Liu, S., Cai, W., Wen, L., & Feng, D. (2012). Semantic-word-based image retrieval for neurodegenerative disorders. *Journal of Nuclear Medicine, 53,* 2309.
60. Liu, S., Cai,W.,Wen, L. & Feng, D. (2011). Volumetric congruent local binary patterns for 3D neurological image retrieval. In P. Delmas & B. Wuensche (Eds.), *The 26th international conference on image and vision computing New Zealand (IVCNZ)* (pp. 272–276). IVCNZ.
61. Liu, S., Liu, S. Q., Pujol, S., Kikinis, R. & Cai, W. (2014). Propagation graph fusion for multi-modal medical content-based retrieval. In *The 13th annual international conference on control, automation, robotics and vision (ICARCV)* (pp. 849–854). IEEE.
62. Liu, S., et al. (2010). A robust volumetric feature extraction approach for 3D neuroimaging retrieval. In *The 32nd annual international conference of the IEEE engineering in medicine and biology society (EMBC)* (pp. 5657–5660). IEEE.
63. Liu, S., et al. (2016). Cross-view neuroimage pattern analysis for Alzheimer's disease staging. *Frontiers in Aging Neuroscience.*
64. Liu, S., et al. (2010). Localized multiscale texture based retrieval of neurological image. In *The 23rd IEEE international symposium on computer-based medical systems (CBMS)* (pp. 243–248). IEEE.
65. Liu, S., et al. (2013). Localized sparse code gradient in Alzheimer's disease staging. In *The 35th annual international conference of the IEEE engineering in medicine and biology society (EMBC)* (pp. 5398–5401). IEEE.
66. Liu, S., et al. (2014). Multi-channel neurodegenerative pattern analysis and its application in Alzheimer's disease characterization. *Computerized Medical Imaging and Graphics, 38,* 436–444. ISSN: 0895-6111.
67. Liu, S., et al. (2013). Multifold bayesian kernelization in Alzheimer's diagnosis. In K. Mori, I. Sakuma, Y. Sato, C. Barillot & N. Navab (Eds.), *The 16th international conference on medical image computing and computer-assisted intervention (MICCAI)* (vol. 8150, pp. 303–310). Berlin, Heidelberg: Springer.
68. Liu, S., et al. (2015). Multimodal neuroimaging computing: A review of the applications in neuropsychiatric disorders. *Brain Informatics, 2,* 167–180.
69. Liu, S., et al. (2015). Subject-centered multi-view neuroimaging analysis. In *The 22nd IEEE international conference on image processing (ICIP).* IEEE.
70. Liu, S., et al. (2011). Localized functional neuroimaging retrieval using 3D discrete curvelet transform. In *IEEE international symposium on biomedical imaging: From nano to macro (ISBI)* (pp. 1877–1880). IEEE.
71. Liu, S., et al. (2015). Multimodal neuroimaging computing: The workflows, methods and platforms. *Brain Informatics, 2,* 181–195.
72. Liu, X., Lai, Y., Wang, X., Hao, C., et al. (2014). A combined DTI and structural MRI study in medicated-naive chronic schizophrenia. *Magnetic Resonance Imaging, 32,* 1–8.
73. Liu, Z., Ding, L., & He, B. (2006). Integration of EEG/MEG with MRI and fMRI in functional neuroimaging. *IEEE Engineering in Medicine and Biology Magazine, 25,* 46–53.
74. Louapre, C., Perlbarg, V., Garcia-Lorenzo, D., Urbanski, M., et al. (2014). Brain networks disconnection in early multiple sclerosis cognitive deficits: An anatomofunctional study. *Human Brain Mapping, 35,* 4706–4717.
75. Maier-Hein, K. H., et al. (2014). Widespread white matter degeneration preceding the onset of dementia. *Alzheimer's & Dementia, S1552–5260,* 1–9.
76. Mazziotta, J., et al. (2001). A probabilistic atlas and reference system for the human brain: International consortium for brain mapping (ICBM). *Philosophical Transactions of the Royal Society of London. Series B: Biological Sciences, 356,* 1293–1322.
77. Medhi, B., Misra, S., Kumar, P., Kumar, P., & Singh, B. (2014). Role of neuroimaging in drug development. *Reviews in the Neurosciences, 25,* 663–673.
78. Modat, M., Simpson, I., Cardoso, M., Cash, D., et al. (2014). *Simulating neurodegeneration through longitudinal population analysis of structural and diffusion weighted MRI data in medical image computing and computer-assisted intervention (MICCAI)* (Vol. 8675). Berlin, Heidelberg: Springer.

79. Mori, S., & van Ziji, P. C. (2002). Fiber tracking: Principles and strategies - A technical review. *NMR in Biomedicine, 15*, 468–480.

80. Morioka, H., Kanemura, A., Morimoto, S., Yoshioka, T., et al. (2013). Decoding spatial attention by using cortical currents estimated from electroencephalography with near-infrared spectroscopy prior information. *NeuroImage, 90*, 128–139.

81. Mueller, S., Keeser, D., Samson, A. C., Kirsch, V., Blautzik, J., et al. (2013). Convergent findings of altered functional and structural briain connectivity in individuals with high functioning autism: A multimodal MRI study. *PLoS ONE, 8*(e67329), 31.

82. Murphy, M. A., O'Brien, T. J., Morris, K., & Cook, M. J. (2004). Multimodality image-guided surgery for the treatment of medically refractory epilepsy. *Journal of Neurosurgery, 100*, 452–462.

83. Murray, C., Abraham, J., Ali, M., Alvarado, M., Atkinson, C., et al. (2013). The state of US health, 1990–2010: Burden of diseases, injuries, and risk factors. *The Journal of the American Medical Association, 310*, 591–608.

84. Nettiksimmons, J., DeCarli, C., Susan Landau, & Beckett, L. (2014). Biological heterogeneity in ADNI amnestic mild cognitive impairment. *Alzheimer's & Dementia, 10*, 511–521.

85. Neuner, I., Kaffanke, J. B., Langen, K.-J., Kops, E. R., Tellmann, L., et al. (2012). Multi-modal imaging utilising integrated MR-PET for human brain tumor assessment. *European Radiology, 22*, 2568–2580.

86. Nguyen, V. T., & Cunnington, R. (2014). The superior temporal sulcus and the N170 during face processing: Single trial analysis of concurrent EEG-fMRI. *NeuroImage, 86*, 492–502.

87. NIH. BRAIN 2025 - A Scientific Vision 2014.

88. Nir, T. M., et al. (2013). Effectiveness of regional DTI measures in distinguishing Alzheimer's disease, MCI, and normal aging. *NeuroImage: Clinical, 3*, 180–195. ISSN: 2213-1582.

89. O'Donnell, L. J., Golby, A. J., & Westin, C.-F. (2013). Fiber clustering versus the parcellation-based connectome. *NeuroImage, 80*, 283–289.

90. Okamoto, M., Dan, K., Shimizu, K., Takeo, K., et al. (2004). Multimodal assessment of cortical activation during apple peeling by NIRS and fMRI. *NeuroImage, 21*, 1275–1288.

91. Okamura, N., Furumoto, S., Harada, R., Tago, T., et al. (2013). Novel 18F-labeled arylquino-line derivatives for noninvasive imaging of Tau pathology in Alzheimer disease. *Journal of Nuclear Medicine, 54*, 1420–1427.

92. Phillips, M. L., & Swartz, H. A. (2014). A critical appraisal of neuroimaging studies of bipolar disorder: Toward a new conceptualization of underlying neural circuitry and a road map for future research. *The American Journal of Psychiatry, 171*, 829–843.

93. Pomarol-Clotet, E., Canales-Rodriguez, E., Salvador, R., Sarro, S., et al. (2010). Medial pre-frontal cortex pathology in schizophrenia as revealed by convergent findings from multimodal imaging. *Molecular Psychiatry, 15*, 823–830.

94. Racine, A. M., Adluru, N., Alexander, A. L., Christian, B. T., et al. (2014). Associations between white matter microstructure and amyloid burden in precinical Alzheimer's disease: A multmodal imaging investigation. *NeuroImage: Clinical, 4*, 604–614.

95. Radua, J., Grau, M., van den Heuvel, O. A., de Schotten, M. T., et al. (2014). Multimodal voxel-based meta-analysis of white matter abnormalities in obsessive- compulsive disorder. *Neuropsychopharmacology, 39*, 1547–1557.

96. Rinck, P. (2014). Magnetic resonance: A critical peer-reviewed introduction in magnetic resonance in medicine. The basic textbook of the European magnetic resonance forum. Chap. 21.

97. Risacher, S. L., et al. (2009). Baseline MRI predictors of conversion from MCI to probable AD in the ADNI cohort. *Current Alzheimer's Research, 6*, 347–361. ISSN: 1875-5828.

98. Rodionov, R., Vollmar, C., Nowell, M., Miserocchi, A., et al. (2013). Feasibility of multimodal 3D neuroimaging to guide implantation of intracranial EEG electrodes. *Epilepsy Research, 107*, 91–100.

99. Rydberg, J., Hammond, C., Grimm, R., Erickson, B., Jack, C. J., et al. (1994). Initial clinical experience in MR imaging of the brain with a fast fluid-attenuated inversion-recovery pulse sequence. *Radiology, 193*, 173–180.

100. Savadjiev, P., Kindlemann, G., Bouix, S., Sheton, M., & Westin, C. (2010). Local white matter geometry from diffusion tensor gradients. *NeuroImage, 49*, 3175–3186.
101. Savadjiev, P., Rathi, Y., Bouix, S., Smith, A. R., et al. (2014). Fusion of white and gray matter geometry: A framework for investigating brain development. *Medical Image Analysis, 18*, 1349–1360.
102. Shah, N. J., Oros-Peusquens, A.-M., Arrbula, J., Zhang, K., Warbrick, T., et al. (2013). Advances in multimodal neuroimaging: Hybrid MR-PET and MR-PET-EEG at 3 T and 9.4 T. *Journal of Magnetic Resonance, 229*, 101–115.
103. Shenton, M., Kubicki, M., & Makris, N. (2014). Understanding alterations in brain connectivity in attention-deficit/hyperactivity disorder using imaging connectomics. *Biological Psychiatry, 76*, 601–602.
104. Sokoloff, L., Reivich, M., Kennedy, C., Des-Rosiers, M., et al. (1977). The [14C]Deoxy-glucose method for the measurement of local cerebral glucose utilization: Theory, procedure and normal values in the consicious and anesthetized albino rat. *Journal of Neurochemistry, 28*, 897–916.
105. Stigler, K. A., McDonald, B. C., Anand, A., et al. (2011). Structural and functional megnetic resonance imaging of autism spectrum disorders. *Brain Research, 1380*, 146–161.
106. Sui, J., Pearlson, G. D., Caprihan, A., Adali, T., Kiehl, K. A., et al. (2011). Discriminating schizophrenia and bipolar disorder by fusing fMRI and DTI in a multimodal CCA+ joint ICA model. *NeuroImage, 57*, 839–855.
107. Taylor, S. F., Stern, E. R., & Gehring, W. J. (2007). Neural systems for error monitoring - recent findings and theoretical perspectives. *The Neuroscientist, 13*, 160–172.
108. Tempany, C. M., Jayender, J., Kapur, T., Bueno, R., et al. (2014). Multimodal imaging for improved diagnosis and treatment of cancers. *Cancer, 121*, 817–827.
109. Tona, F., Petsas, N., Sbardella, E., Prosperini, L., et al. (2014). Multiple sclerosis: Altered thalamic resting-state functional connectivity and its effect on cognitive function. *Radiology, 271*, 814–821.
110. Tong, E., Hou, Q., FFiebach, J. B., & Wintermark, M. (2014). The role of imaging in acute ischemic stroke. *Neurosurgical Focus, 36*, E3.
111. Tournier, J. D., Calamante, F., & Connelly, A. (2007). Robust determination of the fiber orientation distribution in diffusion MRI: Non-negativity constrained super- resolved spherical deconvolution. *NeuroImage, 35*, 1459–1472.
112. Townsend, D. W. (2001). A combined PET/CT scanner: The choices. *Journal of Nuclear Medicine, 42*, 533–534.
113. Tuch, D. S. (2004). Q-ball imaging. *Magnetic Resonance in Medicine, 52*, 1358–1372.
114. Turken, A. U., Herron, T. J., Kang, X., Sorenson, D. J., O'Connor, L. E., et al. (2009). Multimodal surface-based morphometry reveals diffuse cortical atrophy in traumatic brain injury. *BMC Medical Imaging, 9*, 111.
115. Tzourio-Mazoyer, N., Landeau, B., Papathanassiou, D., Crivello, F., et al. (2002). Automated anatomical labelling of activations in SPM using a macroscopy anatomical pacellation of the MNI MRI single-subject brain. *NeuroImage, 15*, 273–289.
116. Wedeen, V., Hagmann, P., Tseng, W., Reese, T., & Weisskoff, R. (2005). Mapping complex tissue architecture with diffusion spectrum magnetic resonance imaging. *Magnetic Resonance in Medicine, 54*, 1377–1386.
117. Wee, C.-Y., Yap, P.-T., Zhang, D., Denny, K., et al. (2012). Identification of MCI individuals using structural and functional connectivity networks. *NeuroImage, 59*, 2045–2056.
118. Westin, C.-F., Szczepankiewicz, F., Pasternak, O., Ozarslan, E., Topgaard, D., et al. (2014). Measurement tensors in diffusion MRI: generalizing the concept of diffusion encoding. *Medical image computing and computer-assisted intervention (MICCAI)* (Vol. 8675, pp. 209–216). Switzerland: Springer.
119. Wong, D. F., Tauscher, J., & Grunder, G. (2009). The role of imaging in proof of concept for CNS drug discovery and development. *Neuropsychopharmacology, 34*, 187–203.
120. Yeh, F., Wedeen, V., & Tseng, W. (2010). Generalized Q-sampling imaging (GQI). *IEEE Transactions on Medical Imaging, 29*, 1626–1635.

121. Young, H., Baum, R., Cremerius, U., Herholz, K., et al. (1999). Measurement of clinical and subclinical tumour reponse using [18F]-Fluorodeoxyglucose and positron emission tomography: Review and 1999 EORTC recommendations. European organization of research and treatment of cancer (EORTC) PET study group. *European Journal of Cancer, 35*, 1773–1782.
122. Zhang, D., Wang, Y., Zhou, L., Yuan, H., & Shen, D. (2011). Multimodal classification of Alzheimer's disease and mild cognitive impairment. *NeuroImage, 55*, 856–867. ISSN: 1053-8119.
123. Zhang, F., et al. (2014). Semantic association for neuroimaging classification of PET images. *Journal of Nuclear Medicine, 55*, 2029.
124. Zhang, W., Arteaga, J., Cashion, D. K., Chen, G., et al. (2013). A highly selective and specific PET tracer for imaging of Tau pathologies. *Journal of Alzheimer's Disease, 31*, 601–612.
125. Zhu, D., Li, K., Terry, D. P., et al. (2014). Connectome-scale assessments of structural and functional connectivity in MCI. *Human Brain Mapping, 35*, 2911–2923.

Chapter 2
Background

This chapter reviews the recent neuroimaging studies with a focus on the characterization of neurodegenerative disorders. These studies fall into four categories based on the primary outputs of these analyses, which correspond to the four layers in the neuroimaging computing architecture, as illustrated in Fig. 2.1. These four layers include the data computing layer, the feature representation layer, the pattern analysis layer and the application development layer.[1]

This chapter is organized as follows. Section 2.1 introduces the most widely used multimodal neuroimaging databases and software packages for large-scale multimodal neuroimaging studies. Section 2.2 reviews the imaging-based features that are used to encode the brain structural and functional changes caused by neurodegeneration, e.g., atrophy and hypo-metabolism. Section 2.3 discusses the current pattern analysis approaches used for capturing the patterns of neurodegenerative pathologies. Section 2.4 demonstrates supervised and unsupervised models in translational applications. As pointed out in Chap. 1, Sect. 1.1.2, MRI and PET are dominantly used to capture the neurodegenerative changes in brain structure and function. Therefore, this chapter focuses on these two neuroimaging modalities and the findings from them.

2.1 Data Computing Layer

2.1.1 Public Neuroimaging Databases

Alzheimer's Disease Neuroimaging Initiative (ADNI) database is the largest database in neurodegeneration research so far, and its volume is still growing. ADNI was

---

[1] Some content of this chapter has been reproduced with permission from [10, 39].

© Springer Nature Singapore Pte Ltd. 2017
S. Liu, *Multimodal Neuroimaging Computing for the Characterization of Neurodegenerative Disorders*, Springer Theses, DOI 10.1007/978-981-10-3533-3_2

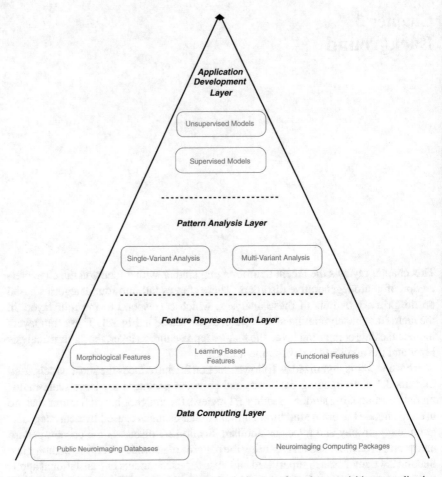

Fig. 2.1 The four layers of neuroimaging analysis architecture, from data acquisition to application development. The higher layers are based on the lower layers

launched in 2003 by the National Institute on Aging (NIA), the National Institute of Biomedical Imaging and Bioengineering (NIBIB), the Food and Drug Administration (FDA), private pharmaceutical companies and non-profit organizations, as a $60 million, 5-year public-private partnership. The primary goal of ADNI has been to test whether serial MRI, PET, other biological markers, and clinical and neuropsychological assessment can be combined to measure the progression of MCI and early AD. Determination of sensitive and specific markers of very early AD progression is intended to aid researchers and clinicians in developing new treatments and monitor their effectiveness, as well as lessen the time and cost of clinical trials. The Principal Investigator of this initiative is Michael W. Weiner, MD, VA Medical Center and University of California - San Francisco. ADNI is the result of the efforts of many co-investigators from a broad range of academic institutions and private corporations, and subjects have been recruited from over 50 sites across the U.S. and Canada. The initial goal of ADNI was to recruit 800 adults, ages 55–90, to participate in the research, approximately 200 cognitively normal older individuals to be followed for 3 years, 400 people with MCI to be followed for 3 years and 200 people with early AD to be followed for 2 years.[2]

The Australia Imaging, Biomarker and Lifestyle Flagship Study of Aging (AIBL) database is a sister database of ADNI. AIBL aims to discover the biomarker, cognitive characteristics, and health and lifestyle factors that may lead to future development of AD, with a particular focus on early detection, towards lifestyle interventions. It was launched in 2006 as one of largest studies of its kind in Australia. AIBL studies are carried out in four research streams, i.e., the clinical and cognitive research stream, lifestyle research stream, biomarker research stream, and neuroimaging research stream. This database contains the datasets acquired from more than 1,000 participants, including patients diagnosed with AD or MCI and age- and sex-matched healthy volunteers.[3]

The Open Access Series of Imaging Studies (OASIS) is an older project compared to ADNI and AIBL. OASIS aims at making MRI data sets of the brain freely available to the scientific community. By compiling and freely distributing MRI data sets, we hope to facilitate future discoveries in basic and clinical neuroscience. OASIS is made available by the Washington University Alzheimer's Disease Research Center, Dr. Randy Buckner at the Howard Hughes Medical Institute (HHMI) at Harvard University, the Neuroinformatics Research Group (NRG) at Washington University School of Medicine, and the Biomedical Informatics Research Network (BIRN). Two collections of MRI databases are available in OASIS. The cross-sectional collection contains 416 subjects including young, middle aged, non-demented and demented older adults. The longitudinal collection has 150 subjects including both non-demented and demented older adults. Each subject was scanned on at least two visits.[4]

2.1.2 Neuroimaging Computing Packages

The sMRI computing workflows usually involve artifact correction, skull stripping, segmentation, registration, surface reconstruction [26], and followed by brain morphometry. The most widely used software packages for sMRI computing include FreeSurfer [18], SPM [67], FSL [31] and ANTs [2], BrainVisa [26], and BrainSuite [73].

PET is exquisitely sensitive for detecting the targeted molecules, an attribute conferred upon it by the choice of radiotracers, such as *2-[^{18}F]fluoro-2-deoxy-D-*

---

[2]For up-to-date information about ADNI, please see www.adni-info.org.

[3]For up-to-date information about AIBL, please refer to http://aibl.csiro.au.

[4]The detailed description of the OASIS projects can be found at http://www.oasis-brains.org/app/template/Index.vm.

glucose (FDG) for the detecting glucose metabolism. PET computing involves serial alignment, spatial normalization and smoothing. SPM and Neurostat[5] packages are available for PET analysis.

FreeSurfer [18], SPM [67], FSL [31] and ANTs [2] are also the common software packages for multimodal neuroimaging computing. 3D Slicer [22, 32] is the mot widely used platform for subject-specific image computing, image editing, visualization and data management. The detailed description of the neuroimaging computing methods and software packages is given in our previous papers [56, 59].

2.2   Feature Representation Layer

2.2.1   Brain Morphological Features

Morphological features, as suggested by the name, aim to model the morphological changes on the brain. They are mainly extracted from sMRI, and can be classified into three groups, the voxel-based features, surface-based features, and ROI-based features.

Voxel-based morphometry (VBM) is a registration-based method used to compute the focal differences between two images, e.g., two scans of the same subjects, two scans of different subjects, or one image and one template. It has been integrated into SPM and FSL packages as a standard method for voxel-wise comparison. It is also referred to as tensor-based morphometry (TBM) when the analysis is carried out in the deformation field. Various voxel-based features can be extracted by VBM and TBM, such as the grey matter density (GMD) derived from the original brain image after registration to a labeled template, or the change rate derived from the registration field based on the comparison of two longitudinal scans [1].

The surface-based features are based on the computing of the brain white matter surface and pial surface reconstructed from the tissue segments. FreeSurfer [18] is a well-established tool for brain tissue segmentation and surface reconstruction. Various measures can be derived from the white matter and pial surfaces, including cortical thickness [23], local gyrification index (LGI) [71], curvedness and shape index [3, 14]. These surface-based features have been extensively reviewed by Mangin et al. [60].

The brain volume can be parcellated into different regions of interest (ROIs) based on a brain template. Popular brain templates include the International Consortium for Brain Mapping (ICBM) template [61] and the Automated Anatomical Labeling (AAL) template [81] in the Montreal Neurological Institute (MNI) coordinates [25]. Grey matter volume (GMV) [28] is the most widely used ROI-feature in neuroimaging analysis. Many other features are also proposed to capture the shape [49, 53] or texture [10, 42, 45, 48, 52, 58, 89] of the ROIs.

---

[5]http://128.208.140.75/Download.

2.2.2 Brain Functional Features

The PET features can reflect particular biochemical processes pertaining to the radioactive tracers. For instance, FDG is able to label the glucose metabolism, and has been a standard tracer in brain tumor studies. The percentage of FDG-PET in brain studies has decreased in recent years due to the introduction of new tracers, such as ^{18}F-BAY94-9172, ^{11}C-SB-13, ^{11}C-BF-227, ^{18}F-AV-45 and ^{11}C-*Pittsburgh compound B* (^{11}C-PiB), which have been reported as tracers for imaging the amyloid plaques in Alzheimer's brains [13, 65, 68, 80].

Cerebral metabolic rate of glucose consumption (CMRGlc) [10, 77] is the most widely used FDG feature at the voxel-level. Recently, Chen proposed the hypo-metabolic convergence index (HCI) to evaluate the hypo-metabolism of the whole-brain, and based on the same principle, he further proposed the amyloid convergence index (ACI) to estimate the amyloid burden on brain [16]. There are also many PET features generic to different tracers, such as the standard uptake value (SUV) [17, 34], mean index [4], z-scores [62], and difference-of-Gaussian (DoG) parametric maps [11, 50].

2.2.3 Learning-Based Feature

Recently, machine learning techniques have been increasingly used in feature extraction from the neuroimaging data. There are many benefits of the learning-based features compared to the hand-engineered features, e.g., they are purely based on the training datasets, but not relying on domain knowledge of the disorders or imaging modalities. Deep learning, in particularly, uses a end-to-end learning strategy, which is suitable for learning the high-level and multimodal features. In addition, as the datasets are growing rapidly in volume, machine learning models can further benefit from such increasingly large-scale datasets [5].

Recently, Brosch and Tam reported that deep learning is effective in capturing the shape variations of brain MRI that highly correlated with disease pathologies, such as the enlarged ventricles [6]. In another study, Suk et al. [79] proposed a deep learning approach for extracting high-level features from low-level multimodal neuroimaging features. They trained a stacked auto-encoder (SAE) for each imaging modality, then then combined the learnt high-level features using a multi-kernel support vector machine (MK-SVM). Recently, Suk et al. proposed another deep learning model based on the the deep Boltzmann machine (DBM), which was trained using the 3D patches, instead of the low-level features, from the multimodal neuroiamging data [78].

2.3 Pattern Analysis Layer

Neurodegenerative disorders may progress in certain patterns. In the case of AD, for example, its pathology commonly starts in the hippocampus and entorhinal cortex, then spreads to the temporal lobe and the posterior cingulate, and then to the parietal, prefrontal and orbitofrontal regions, and finally throughout the entire brain [20, 21, 70]. The patterns of pathologies also vary between different neurological disorders. For example, FTD selectively affects the frontal lobe and may extend backward to the temporal lobe, thus its pattern markedly different from that of AD.

2.3.1 Single-Variant Analysis

The simplest approach for pattern analysis is single-variant methods, e.g., t-test for two-class comparison and ANOVA for multiple-class comparison. When these methods are applied to voxel-wise analysis, such as VBM, we may able to derive the t-maps [10] or z-score maps [62]. Figure 2.2 illustrates the CMRGlc maps, functional normalized CMRGlc maps, t-maps and thresholded t-maps of an AD patient and a FTD patient. In this example, the t-maps are generated by comparison of the patient's image to those of the normal controls. Same methods can also be applied to ROI-based analysis, as demonstrated in many previous studies [28, 35, 43, 50, 51, 54].

2.3.2 Multi-Variant Analysis

Since single-variant analysis methods ignore correlation between individual variants in the analysis, multivariant methods, such as lasso and elastic net (EN), are therefore increasingly used in pattern analysis due to their capability of identifying the most important features/regions as well as reducing the redundant features/regions with high correlation. For example, an elastic net logistic regression model was proposed by Shen et al. [75] to select disease-relevant biomarkers and classify AD and MCI. Recently, Zhu et al. proposed a joint regression and classification framework for AD/MCI diagnosis based on lasso [93].

Multi-variate analysis methods are also inherently suitable for multimodal pattern analysis and not restricted to imaging data only. In the two examples given above, Zhu et al. used MRI, PET and Cerebrospinal Fluid (CSF) biomarkers in their study, whereas Shen et al. jointly analyzed of MRI data and proteomic data.

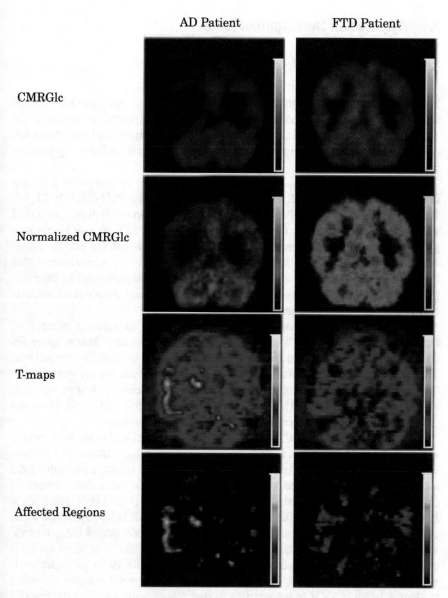

Fig. 2.2 Examples of CMRGlc maps, functional normalized CMRGlc maps, t-maps and thresholded t-maps of an AD patient and a FTD patient. These maps were registered to the MNI coordinates using SPM. Figure reproduced with permission from [10]

2.4   Application Development Layer

2.4.1   Supervised Models

There are a variety of applications developed for neurodegenerative disorder characterization. A substantial proportion of them focus on classification/prediction of the cognitive status. These applications are all based on the supervised models, which require training datasets with ground truth, i.e., the diagnosis confirmed by the doctors.

Most neuroimaging classification/prediction studies are carried out in a similar fashion. The primary features are usually extracted from the MRI data [19, 21, 66, 70, 75, 76, 79, 85] and/or PET data [75, 76, 79, 85], and sometimes combined with other biomarkers, e.g., CSF measures [19, 76, 79, 85], genetic biomarkers [75, 76, 85] and clinical assessments [85]. The features are then fed into the classifiers, which are trained for future classifications. Various classifiers have been used in these studies, predominantly SVM [19, 21, 53, 79, 85], complemented by Bayesian [55], Elastic Net [75], k-nearest-neighbors (kNN) [66], linear discriminant analysis (LDA) [76], and softmax regression [37, 41].

With increasing attention given in the multimodal analysis, a number of multimodal classifiers have been proposed, which create a new feature space for the multimodal features and then train a single model to classify the patients. The most straightforward solution is to concatenate input multi-view features into high-dimensional vectors, and then apply dimension reduction or feature selection approaches, such as t-test [51], ISOMAP [66], Elastic Net [40, 75], or combinations of these methods [35, 54], to reduce the 'curse of dimensionality'.

These concatenation-based methods are simple and effective, but the inter-subject variations cannot be eliminated, since the inter-subject similarity measured by different features vary in scales and variances. With a focus on the subjects, the multimodal analysis advances rapidly due to the research efforts on the multi-view embedding (ME) methods, such as Multi-View Spectral Embedding (MSE) [84], Multi-View Local Linear Embedding (MLLE) [74], Co-Neighbor MSE [15], Supervised MSE [49], which are based on manifold-learning. These methods extract the geometric structures of local patches across multiple feature spaces, and then align the local patches in a unified feature space with maximum preservation of the geometric relationships. Hinrichs et al. [29, 30] and Zhang et al. [86], on the contrary, extended the kernel tricks in SVM to the multiple feature spaces, and combined the features at the classifier level.

2.4.2 Unsupervised Models

The unsupervised models are different from the supervised models in that they do not directly provide a second opinion for the physician, but support clinical decisions in an indirect way. Ground truth is not necessarily needed in these applications.

Medical content-based retrieval (MCBR) is a typical unsupervised application that can provide clinical decision support by retrieving the similar subjects as references for a query. Various content-based neurological image retrieval systems have been reported [7, 9–12, 24, 35, 36, 38, 42–48, 52, 57, 58, 69, 82, 87–92]. These studies mainly focused on single modal data or features, such as High Resolution Computed Tomography (HRCT) [24], Positron Emission Tomography (PET) [9–12, 38, 42, 44, 45, 48, 52, 54, 58, 91], Single Photon Emission Computed Tomography (SPECT) [69], Magnetic Resonance Imaging (MRI) [35, 43, 57, 82], and functional MRI (fMRI) [7]. For example, Cai et al. [9] previously proposed a dynamic brain PET image retrieval system based on pixel-wise tissue time activity curve (TTAC). They further extended their previous work to a volume of interest (VOI)-based retrieval system [33]. Wong et al. [83] established a neuro-informatics database system (NIDS) with co-registered static PET and MR image data, supporting geometric, metabolic and textual features. Batty et al. [4] designed a PET image retrieval system based on a combination of anatomical and functional ROI features for retrieval of demented cases. There is a clear trend of using the bag of visual words (BoVW) model for MCBR. Various visual words and models were proposed [8, 24, 27, 90], mostly based on the low-level features, such as texture, shape, size, intensity or a combination of them. Some studies also tried to add semantic annotations [44, 64, 72, 88, 89, 91] to the medical images.

Recently, the MCBR model has been extended to retrieving longitudinal brain deformations. In one of our recent studies, we developed a MCBR model to simulate the future brain development, based on both the longitudinal and cross-sectional information. It was assumed that brains with similar morphological deformations at both previous and current time points will have similar future development [39, 63]. Figure 2.3a–c illustrates the longitudinal images and the deformation field of registration; (d–f) shows an example of the query (deformation field) and the retrieval results with the most similar longitudinal changes. As the database is growing in volume, we may expect better performance with the increasing large-scale database.

Another extension of the current MCBR framework is multimodal MCBR methods, which take the benefits from the the emerging multimodal medical data with complementary information. For example, a bag of semantic words (BoSW) model based on deep learning was recently proposed for multimodal MCBR. A set of low-level features were extracted from the multimodal medical imaging data and then translated to the symptom severity degrees by clinical symptom quantization. Finally, high-level semantic words were deduced by learning the patterns of the symptoms [47].

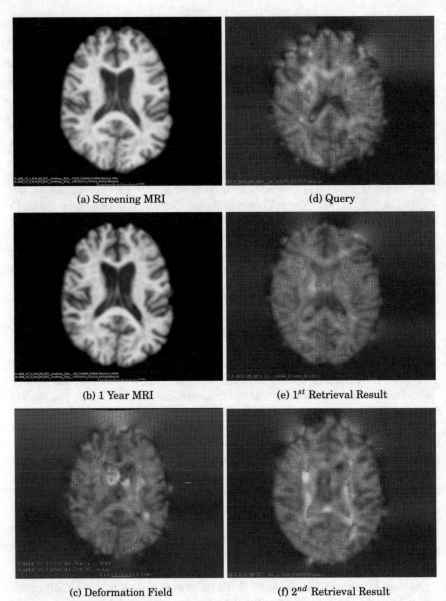

(a) Screening MRI   (d) Query

(b) 1 Year MRI   (e) 1st Retrieval Result

(c) Deformation Field   (f) 2nd Retrieval Result

Fig. 2.3 *First column* an example of the longitudinal screening scan (**a**) the follow up scan in 1 year (**b**), and their deformation field (**c**) after registration. *Second column* an example of deformation-based query (**d**) and the two top ranked retrieval results (**e, f**) with the most similar longitudinal changes. Figure reproduced with permission from [39]

References

1. Ashburner, J., & Friston, J. K. (2000). Voxel-based morphometry - the methods. *NeuroImage, 11,* 805–821.
2. Avants, B. B., et al. (2010). A reproducible evaluation of ANTs similairty metric performance in brain image registration. *NeuroImage, 54,* 2033–2044.
3. Awate, S. P., Yushkevich, P. A., Song, Z., Licht, D. J., & Gee, J. C. (2010). Cerebral cortical folding analysis with multivariate modeling and testing: Studies on gender differences and neonatal development. *NeuroImage, 53,* 450–459. ISSN: 1053-8119.
4. Batty, S., Clark, J., Fryer, T., & Gao, X. (2008). Prototype system for semantic retrieval of neurological PET images English. In X. Gao, H. Müller, M. Loomes, R. Comley, & S. Luo (Eds.), *Medical imaging and informatics* (Vol. 4987, pp. 179–188). Berlin: Springer. ISBN: 978-3-540-79489-9.
5. Bengio, Y., Courville, A., & Vincent, P. (2013). Representation learning: A review and new perspectives. *IEEE Transactions on Pattern Analysis and Machine Intelligence, 35,* 1798–1828.
6. Brosch, T., & Tam, R. (2013). *Manifold learning of brain MRIs by deep learning in medical image computing and computer-assisted intervention (MICCAI).* Berlin: Springer.
7. Buckner, R. L., Koutstaal, W., Schacter, D. L., Wagner, A. D., & Rosen, B. R. (1998). Functional-anatomic study of episodic retrieval using fMRI: I. Retrieval effort versus retrieval success. *NeuroImage,7,* 151–162. ISSN: 1053-8119.
8. Burner, A., Donner, R., Mayerhoefer, M., Hozer, M., et al. (2011). *Texture bags: Anomaly retrieval in the medical images based on local 3D-texture similarity in the MICCAI workshop on medical content-based retrieval for clinical decision support (MCBR-CDS).* Berlin: Springer.
9. Cai, W., Feng, D., & Fulton, R. (2000). Content-based retrieval of dynamic PET functional images. *IEEE Transactions on Information Technology in Biomedicine, 4,* 152–158.
10. Cai, W., et al. (2010). *3D neurological image retrieval with localized pathology-centric CMR-Glc patterns in the 17th IEEE international conference on image processing (ICIP)* (IEEE, 2010) (pp. 3201–3204).
11. Cai, W., et al. (2014). *A 3D difference of gaussian based lesion detector for brain PET in IEEE international symposium on biomedical imaging: From nano to macro (ISBI)* (IEEE, 2014) (pp. 677–680).
12. Cai, W., et al. (2014). *Automated feedback extraction for medical imaging retrieval in IEEE international symposium on biomedical imaging: From nano to macro (ISBI)* (IEEE, 2014) (pp. 907–910).
13. Carpenter, A. J., Pontecorvo, M., Hefti, F., & Skovronsky, D. (2009). The use of the exploratory IND in the evaluation and development of 18F-PET radiopharmaceuticals for amyloid imaging in the brain: A review of one company's experience. *The Quarterly Journal of Nuclear Medcine and Molecular Imaging, 53,* 387–393.
14. Cash, D. M., et al. (2012). In N. Ayache, H. Delingette, P. Golland, & K. Mori (Eds.), *Cortical folding analysis on patients with Alzheimer's disease and mild cognitive impairment in medical image computing and computer-assisted intervention (MICCAI)* (Vol. 7512, pp. 289–296). Berlin: Springer. ISBN: 978-3-642-33453-5.
15. Che, H., et al. (2014). *Co-neighbor multi-view spectral embedding for medical contentbased retrieval in IEEE international symposium on biomedical imaging: From nano to macro (ISBI)* (IEEE, 2014) (pp. 911–914).
16. Chen, K., Ayutyanont, N., Langbaum, J. B., Fleisher, A. S., Reschke, C., et al. (2011). Characterizing Alzheimer's disease using a hypometabolic convergence index. *NeuroImage, 56,* 52–60. ISSN: 1053-8119.
17. Clark, C. M., et al. (2012). Cerebral PET with florbetapir compared with neuropathology at autopsy for detection of neuritic amyloid-OE≤ plaques: A prospective cohort study. *The Lancet Neurology, 11,* 669–678. ISSN: 1474-4422.
18. Dale, A. M., Fischl, B., & Sereno, M. I. (1999). Cortical surface-based analysis: I. Segmentation and surface reconstruction. *NeuroImage, 9,* 179–194.

19. Davatzikos, C., Bhatt, P., Shaw, L., Batmanghelich, K., & Trojanowski, J. (2011). Prediction of MCI to AD conversion, via MRI, CSF biomarkers, pattern classification. *Neurobiology of Aging, 32*, 2322.e19-e27.

20. Desikan, R., Cabral, H., Hess, C., Dilon, W., et al. (2009). Automated MRI measures identify individuals with mild cognitive impairment and Alzheimer's disease. *BRAIN, 132*, 2048–2057.

21. Fan, Y., Batmanghelich, N., Clark, C., & Davatzikos, C. (2008). Spatial patterns of brain atrophy in MCI patients, identified vis high-dimensional pattern classificaiton, predic subsequent cognitie decline. *NeuroImage, 39*, 1731–1743.

22. Fedorov, A., et al. (2012). 3D Slicer as an image computing platform for the quantitative imaging network. *Magnetic Resonance Imaging, 30*, 1323–1341. ISSN: 0730-725X.

23. Fischl, B., & Dale, A. M. (2000). Measuring the thickness of the human cerebral cortex from magnetic resonance images. *Proceedings of the National Academy of Sciences, 97*, 11050–11055.

24. Foncubierta-Rodríguez, A., Depeursinge, A., & Müller, H. (2011). In H. Müller, H. Greenspan, & T. Syeda-Mahmood (Eds.), *Using multiscale visual words for lung texture classification and retrieval in the MICCAI workshop on medical content-based retrieval for clinical decision support* (Vol. 7075, pp. 69–79). Berlin: Springer. ISBN: 978-3-642-28459-5.

25. Fonov, V., Evans, A., Botteron, K., Almli, C., et al. (2010). Unbiased average age-approapriate atlases for pediatric studies. *NeuroImage, 54*, 313–327.

26. Geffroy, D., et al. (2011). *BrainVISA: A complete software platform for neuroimaging in python in neuroscience workshop.*

27. Hass, S., Donner, R., Holzer, A., et al. (2011). *Superpixel-based interest points for effective bags of visual words medical image retrieval in the MICCAI workshop on medical content-based retrieval for clinical decision support.* Berlin: Springer.

28. Heckemann, R. A., et al. (2011). Automatic morphometry in alzheimer's disease and mild cognitive impairment. *NeuroImage, 56*, 2024–2037. ISSN: 1053-8119.

29. Hinrichs, C., Singh, V., Xu, G., & Johnson, S. (2011). Predictive markers for AD in a multi-modality framework: An analysis of MCI progression in the ADNI population. *NeuroImage, 55*, 574–589.

30. Hinrichs, C., Singh, V., Xu, G., & Johnson, S. (2009). In G. Yang (Ed.), *MKL for robust multi-modality AD classification in medical image computing and computer-assisted intervention (MICCAI)* (Vol. 5762, pp. 786–794). Berlin: Springer.

31. Jenkinson, M., Beckmann, C., Behrens, T., Woolrich, M., & Smith, S. (2012). FSL. *NeuroImage, 62*, 782–790.

32. Kikinis, R., Pieper, S. D., & Vosburgh, K. (2014). 3D Slicer: A platform for subject-specific image analysis, visualization, and clinical support. In F. A. Jolesz (Ed.), *Intraoperative imaging and image-guided therapy* (Vol. 3, pp. 277–289). Berlin: Springer.

33. Kim, J., Cai, W., Feng, D., & Wu, H. (2006). A new way for multi-dimentional medical data management: Volume of interest (VOI)-based retrieval of medical images with visual and functional features. *IEEE Transactions on Information Technology in Biomedicine, 10*, 598–607.

34. Landau, S. M., et al. (2013). Comparing positron emission tomography imaging and cerebrospinal fluid measurements of beta-amyloid. *Annals of Neurology, 74*, 826–836. ISSN: 1531-8249.

35. Liu, S., Cai, W., Wen, L., & Feng, D. (2013). *Multi-channel brain atrophy pattern analysis in neuroimaging retrieval in IEEE international symposium on biomedical imaging: From nano to macro (ISBI)* (IEEE, 2013) (pp. 206–209).

36. Liu, S. Q., et al. (2015). *Content-based retrieval of brain diffusion magnetic resonance image in multimodal retrieval in the medical domain*, (Vol. 9059). Berlin: Springer.

37. Liu, S. Q., et al. (2014). *Early diagnosis of Alzheimer's disease with deep learning in IEEE international symposium on biomedical imaging: from nano to macro (ISBI)* (IEEE, 2014) (pp. 1015–1018).

38. Liu, S. Q., et al. (2014). High-level feature based PET image retrieval with deep learning architecture. *Journal of Nuclear Medicine, 55*, 2018.

39. Liu, S. Q., et al. (2015). Longitudinal brain MR retrieval with diffeomorphic demons registration: What happened to those patients with similar changes? In *IEEE international symposium on biomedical imaging: from nano to macro (ISBI)* (IEEE, 2015) (pp. 588–591).
40. Liu, S. Q., et al. (2015). Multi-modal neuroimaging feature learning for multi-class diagnosis of Alzheimer's disease. *IEEE Transactions on Biomedical Engineering, 62*, 1132–1140.
41. Liu, S. Q., et al. (2015). *Multi-phase feature representation learning for neurodegenerative disease diagnosis in the 1st Australian conference on artificial life and computational intelligence (ACALCI)* (pp. 350–359) Berlin: Springer.
42. Liu, S., Cai, W., Wen, L., & Feng, D. (2012). *Multiscale and multiorientation feature extraction with degenerative patterns for 3D neuroimaging retrieval in the 19th IEEE international conference on image processing (ICIP)* (IEEE, 2012) (pp. 1249–1252).
43. Liu, S., Cai, W., Wen, L., & Feng, D. (2013). *Neuroimaging biomarker based prediction of Alzheimer's disease severity with optimized graph construction in IEEE international symposium on biomedical imaging: From nano to macro (ISBI)* (IEEE, 2013) (pp. 1324–1327).
44. Liu, S., Cai, W., Wen, L., & Feng, D. (2012). Semantic-word-based image retrieval for neurodegenerative disorders. *Journal of Nuclear Medicine, 53*, 2309.
45. Liu, S., Cai, W., Wen, L., & Feng, D. (2011). Volumetric congruent local binary patterns for 3D neurological image retrieval. In P. Delmas, B. Wuensche, & J. James (Eds.), *The 26th international conference on image and vision computing New Zealand (IVCNZ)* (IVCNZ, 2011) (pp. 272–276).
46. Liu, S., Liu, S. Q., Pujol, S., Kikinis, R., & Cai, W. (2014). Propagation graph fusion for multi-modal medical content-based retrieval. In *The 13th annual international conference on control, automation, robotics and vision (ICARCV)* (IEEE, 2014) (pp. 849–854).
47. Liu, S., et al. (2013). A bag of semantic words model for medical content-based retrieval In T. Syeda-Mohmood, H. Greenspan, & A. Madahushi (Eds.), *The MICCAI workshop on medical content-based retrieval for clinical decision support (MCBR-CDS)* (pp. 1–8). IBM Press.
48. Liu, S., et al. (2010). A robust volumetric feature extraction approach for 3D neuroimaging retrieval In *The 32nd annual international conference of the IEEE engineering in medicine and biology society (EMBC)* (IEEE, 2010) (pp. 5657–5660).
49. Liu, S., et al. (2013). A supervised multiview spectral embedding method for neuroimaging classification In *The 20th IEEE international conference on image processing (ICIP)* (IEEE, 2013) (pp. 601–605).
50. Liu, S., et al. (2016). Cross-view neuroimage pattern analysis for Alzheimer's disease staging. *Frontiers in Aging Neuroscience.*
51. Liu, S., et al. (2011). Generalized regional disorder-sensitive-weighting scheme for 3D neuroimaging retrieval. In *The 33rd annual international conference of the IEEE engineering in medicine and biology society (EMBC)* (IEEE, 2011) (pp. 7009–7012).
52. Liu, S., et al. (2010). Localized multiscale texture based retrieval of neurological image. In *The 23rd IEEE international symposium on computer-based medical systems (CBMS)* (IEEE, 2010) (pp. 243–248).
53. Liu, S., et al. (2013). Localized sparse code gradient in Alzheimer's disease staging. In *The 35th annual international conference of the IEEE engineering in medicine and biology society (EMBC)* (IEEE, 2013) (pp. 5398–5401).
54. Liu, S., et al. (2014). Multi-channel neurodegenerative pattern analysis and its application in Alzheimer's disease characterization. *Computerized Medical Imaging and Graphics, 38*, 436–444. ISSN: 0895-6111.
55. Liu, S., et al. (2013). Multifold Bayesian kernelization in Alzheimer's diagnosis. In K. Mori, I. Sakuma, Y. Sato, C., Barillot & N. Navab (Eds.), *The 16th international conference on medical image computing and computer-assisted intervention (MICCAI)* (Vol. 8150, pp. 303–310). Berlin: Springer.
56. Liu, S., et al. (2015). Multimodal neuroimaging computing: A review of the applications in neuropsychiatric disorders. *Brain Informatics, 2*, 167–180.
57. Liu, S., et al. (2015). Subject-centered multi-view neuroimaging analysis. In *The 22nd IEEE international conference on image processing (ICIP)* (IEEE, 2015).

58. Liu, S., et al. (2011). Localized functional neuroimaging retrieval using 3D discrete curvelet transform. In *IEEE international symposium on biomedical imaging: From nano to macro (ISBI)* (IEEE, 2011) (pp. 1877–1880).
59. Liu, S., et al. (2015). Multimodal neuroimaging computing: The workflows, methods and platforms. *Brain Informatics, 2*, 181–195.
60. Mangin, J., Jouvent, E., & Cachia, A. (2010). In-vivo measurement of cortical morphology: Means and meanings. *Current Opinion in Neurology, 23*, 359–367.
61. Mazziotta, J., et al. (2001). A probabilistic atlas and reference system for the human brain: International consortium for brain mapping (ICBM). Philosophical Transactions of the Royal Society of London. Series B: Biological Sciences, *356*, 1293–1322.
62. Minoshima, S., Frey, K. A., Koeppe, R. A., Foster, N. L., & Kuhl, D. E. (1995). A diagnostic approach in Alzheimer's disease using three-dimensional stereotactic surface projections of fluorine-18-FDG PET. *Journal of Nuclear Medicine, 36*, 1238–1248.
63. Modat, M., Simpson, I., Cardoso, M., Cash, D., et al. (2014). *Simulating neurodegeneration through longitudinal population analysis of structural and diffusion weighted MRI data in medical image computing and computer-assisted intervention (MICCAI)* (Vol. 8675, pp. 57–64). Berlin: Springer.
64. Moller, M., Sintek, M. (2007). A generic framework for semantic medical image retrieval. In *The international workshop on knowledge acquisition from multimedia content* (Vol. 253).
65. Ni, R., Gillberg, P., Bergfors, A., Marutle, A., & Nordberg, A. (2013). Amyloid tracers detect multiple binding sites in Alzheimer's disease brain tissue. *Brain, 136*, 2217–2227.
66. Park, H. (2012) ISOMAP induced manifold embedding and its application to Alzheimer's disease and mild cognitive impairment. *Neuroscience Letters, 513*, 141–145. ISSN: 0304-3940.
67. Penny, W., Friston, K., Ashbuner, J., Kiebel, S., et al. (2011). *Statistical parametric mapping: The analysis of functional brain images*. New York: Academic Press.
68. Perrin, R. J., Fagan, A. M., & Holtzmann, D. M. (2009). Multimodal techniques for diagnosis and prognosis of Alzheimer's disease. *Nature, 461*, 916–922.
69. Ramírez, J., et al. (2009). In M. Köppen, N. Kasabov, & G. Coghill (Eds.), *Early detection of the Alzheimer disease combining feature selection and kernel machines in advances in neuro-information processing* (pp. 410–417). Berlin: Springer. ISBN: 978-3-642-03039-0.
70. Risacher, S. L., et al. (2009). Baseline MRI predictors of conversion from MCI to probable AD in the ADNI cohort. *Current Alzheimer's Research, 6*, 347–361. ISSN: 1875-5828.
71. Schaer, M., et al. (2008). A surface-based approach to quantify local cortical gyrification. *IEEE Transactions on Medical Imaging, 27*, 161–170.
72. Seifert, S., Thoma, M., Stegmaier, F., et al. (2011). Combined semantic and similarity search in medical image databases. In *SPIE medical imaging* (Vol. 7967).
73. Shattuck, D., & Leahy, R. (2002). BrainSuite: An automated cortical surface identification tool. *Medical Image Analysis, 8*, 129–142.
74. Shen, H., Tao, D., & Ma, D. (2013). Multiview locally linear embedding for effective medical image retrieval. *PLoS ONE, 8*, e82409.
75. Shen, L., et al. (2011). Identifying neuroimaging and proteomic biomarkers for MCI and AD via the elastic net. In T. Liu, D. Shen, L. Ibanez, & X. Tao (Eds.), Multimodal brain image analysis (MBIA) (Vol. 7012, pp. 27–34). Berlin: Springer. ISBN: 978-3-642-24445-2.
76. Singh, N., Wang, A., Sankaranarayanan, P., Fletcher, P. & Joshi, S. (2012). Genetic, structural and functional imaging biomarkers for early detection of conversion from MCI to AD. In N. Ayache, H. Delingette, P. Golland, & K. Mori (Eds.), *Medical image computing and computer-assisted intervention (MICCAI)* (Vol. 7510, pp. 132–140). Berlin: Springer. ISBN: 978-3-642-33414-6.
77. Sokoloff, L., Reivich, M., Kennedy, C., Des-Rosiers, M., et al. (1977). The [14C]deoxy-glucose method for the measurement of local cerebral glucose utilization: Theory, procedure and normal values in the consicious and anesthetized albino rat. *Journal of Neurochemistry, 28*, 897–916.
78. Suk, H., Lee, S., & Shen, D. (2014). Hierarchical feature representation and multimodal fusion with deep learning for AD/MCI diagnosis. *NeuroImage, 101*, 569–582.

79. Suk, H.-I., Lee, S., & Shen, D. (2013). Latent feature representation with stacked auto-encoder for AD/MCI diagnosis. *Brain Structure and Function, 220*, 841–959.
80. Thompson, P. M., Ye, L., Morgenstem, J. L., Sue, L., Beach, T. G., et al. (2009). Interaction of the amyloid imaging tracer FDDNP with hallmark Alzheimer's disease pathologies. *Journal of Neurochemistry, 109*, 623–630.
81. Tzourio-Mazoyer, N., Landeau, B., Papathanassiou, D., Crivello, F., et al. (2002). Automated anatomical labelling of activations in SPM using a macroscopy anatomical pacellation of the MNI MRI single-subject brain. *NeuroImage, 15*, 273–289.
82. Unay, D., Ekin, A., & Jasinschi, R. (2010). Local structure-based region-of-interest retrieval in brain MR images. *IEEE Transactions on Information Technology in Biomedicine, 14*, 897–903.
83. Wong, S., Hoo, K., Gao, X., et al. (2004). A neuroinformatics database system for disease-oriented neuroiamgign research. *Academic Radiology, 11*, 345–358.
84. Xia, T., Tao, D., Mei, T., & Zhang, Y. (2010). Multiview spectral embedding. *IEEE Transactions on Systems, Man, and Cybernetics, Part B: Cybernetics, 40*, 1438–1446.
85. Ye, J., et al. (2012). Sparse learning and stability selection for predicting MCI to AD conversion using baseline ADNI data. *BMC Neurology, 12*, 46. ISSN: 1471-2377.
86. Zhang, D., Wang, Y., Zhou, L., Yuan, H., & Shen, D. (2011). Multimodal classification of Alzheimer's disease and mild cognitive impairment. *NeuroImage, 55*, 856–867. ISSN: 1053-8119.
87. Zhang, F., et al. (2015). Dictionary refinement with visual word significance for medical image retrieval. *Neurocomputing*.
88. Zhang, F., et al. (2014). Latent semantic association analysis for medical image retrieval. In *International conference on digital image computing: techniques and applications (DICTA)* (pp. 1–6).
89. Zhang, F., et al. (2015). Pairwise latent semantic association for similarity computation in medical imaging. *IEEE Transactions on Biomedical Engineering*.
90. Zhang, F., et al. (2015). Ranking-based vocabulary pruning in bag-of-features for image retrieval. In *The 1st Australian conference on artificial life and computational intelligence (ACALCI)* (Vol. 8955, pp. 436–445). Berlin: Springer.
91. Zhang, F., et al. (2014). Semantic association for neuroimaging classification of PET images. *Journal of Nuclear Medicine, 55*, 2029.
92. Zhang, L., et al. (2013). Graph cuts based relevance feedback in image retrieval. In *The 20th IEEE international conference on image processing (ICIP)* (IEEE, 2013) (pp. 4358–4362).
93. Zhu, X., Suk, H.-I., & Shen, D. (2014). A novel matrix-similarity based loss function for joint regression and classification in AD diagnosis. *NeuroImage, 100*, 91–105.

Chapter 3
ADNI Datasets and Pre-processing Protocols

Datasets used in this study were obtained from the Alzheimer's Disease Neuroimaging Initiative (ADNI) baseline database[1] (adni.loni.usc.edu). A brief introduction of the ADNI database has been given in Chap. 2, Sect. 2.1.1. We divided this dataset into four subsets, according to the imaging modalities and the analyses to be carried out. This chapter describes these four ADNI subsets, as well as the selection criteria and pre-processing protocols.[2]

Section 3.1 gives the details of the ADNI MRI subset, which was mainly used to test methods for modeling the brain morphological changes, such as cortical thickness [10], grey matter volume (GMV) [1], curvedness and shape index [2, 7], local gyrification index (lGI) [32], solidity and convexity ratios [25–27], and texture features [5, 22, 28]. Section 3.2 describes the ADNI PET subset, which was used to validate a number of functional feature descriptors, such as standard uptake value (SUV) [9, 19], cerebral metabolic rate of glucose consumption (CMRGlc) [5, 34], hypo-metabolic convergence index (HCI) and amyloid convergence index (ACI) [8], mean index [4], z-scores [30], and difference-of-Gaussian (DoG) parametric maps [6]. Section 3.3 introduces the multimodal MRI-PET subset, which has 331 subjects with both MRI and PET scans. The high-level learning-based feature extraction methods, such as stacked auto-encoder (SAE) [21, 35], were tested using this subset. An additional ADNI dMRI subset is introduced in Sect. 3.4, which was used to demonstrate effectiveness of the proposed propagation graph fusion algorithm in content-based retrieval [20], as described in Chap. 7.

---

[1] As such, the investigators within the ADNI contributed to the design and implementation of ADNI and/or provided data but did not participate in analysis or writing of this report. A complete listing of ADNI investigators can be found at:http://adni.loni.usc.edu/wp-content/uploads/how_to_apply/ ADNI_Acknowledgement_List.pdf.

[2] Some content of this chapter has been reproduced with permission from [20, 23, 24, 26].

© Springer Nature Singapore Pte Ltd. 2017
S. Liu, *Multimodal Neuroimaging Computing for the Characterization
of Neurodegenerative Disorders*, Springer Theses, DOI 10.1007/978-981-10-3533-3_3

3.1   The ADNI MRI Subset

At the beginning of this study, there were 816 subjects recruited in ADNI base-line cohort, each with at least one T1-weighted (1.5 or 3 T) MRI. However, the FDG-PET scans were only obtained from 369 subjects and the dMRI data were available for 233 subjects. Therefore, the ADNI MRI subset contained 816 subjects, the PET and MRI-PET subsets both contained 369 subjects, and the dMRI subset had 233 subjects, before further selection.

These subjects were grouped according to their initial diagnoses, including the Alzheimer's disease (AD), mild cognitive impairment (MCI), and normal control (NC). The MCI group was further divided into two sub-groups: if the MCI subjects converted to AD in half year to 3 years from the screening scan, they were labeled as the MCI converters, denoted by 'cMCI'; other MCI non-converters were labeled as 'ncMCI'.

3.1.1   Screening Criteria

Subjects in the MRI subset were selected using the following recruitment criteria. Twenty subjects with the multiple conversions or reversions were excluded, as multiple conversions or revisions may be due to the inaccurate initial diagnosis. However, we retained only one subject with multiple conversions, for this subject converted from NC through MCI to AD in three years and did not convert back to MCI or NC. Another 21 normal subjects that converted to MCI were excluded, as their cognitive status were not definite and should not be considered as healthy volunteers. Finally, 17 subjects were rejected during pre-processing, since they showed intolerable image distortions during visual examination by the experts. Therefore, totally 758 participants were selected in the ADNI MRI subset, including 180 AD patients, 160 cMCI patients, 214 ncMCI patients and 204 NC subjects.

3.1.2   Pre-processing Protocols

The MRI analysis needs to be carried out in the original brain space to maximize the preservation of the anatomical information. The MRI data in ADNI baseline cohort have been corrected according to the ADNI MRI image correction protocol [14], and further labeled with 83 brain ROIs through the multi-atlas propagation with enhanced registration (MAPER) approach [12, 13], which were based on 30 T1-weighted MRI images acquired from the National Society for Epilepsy at Chalfont, UK., using the protocol described by Hammers et al. [11]. Figure 3.1 shows an example of the

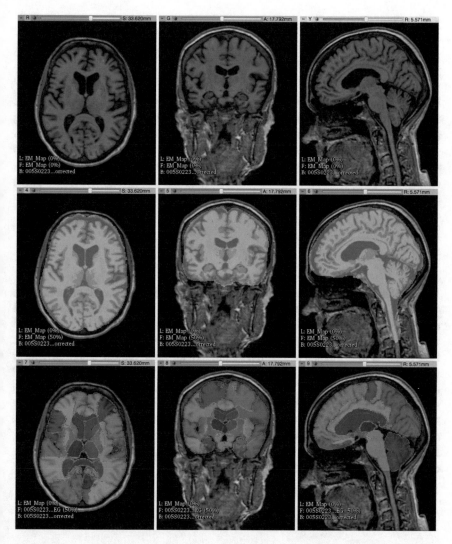

Fig. 3.1 An example of the segmented brain of a cMCI subject. *Upper row*: T1-weighted MRI scan; *middle row*: result of tissue classification; *lower row*: segmentation generated with MAPER [12]

segmentation result of a cMCI patient in the ADNI MRI subset. The label outlines are superimposed on the transverse, coronal and sagittal planes of the MRI scan, as shown in Fig. 3.1. The names of these ROIs are listed in Appendix B.

3.2   ADNI PET Subset

3.2.1   Screening Criteria

In the PET subset, 369 subjects with FDG-PET scans were initially recruited. The analyses on PET data does not involve MRI, but MRI is still needed in pre-processing for providing the anatomical reference. As described in Sect. 3.1, many MRI scans were removed from the MRI subset, therefore 17 subjects in the PET subset with no available MRI scan were excluded, resulting in a downsized database of 352 subjects. These patients were further divided into 3 groups according to their screening diagnoses, i.e., 85 AD patients, 181 MCI patients and 86 NC subjects.

3.2.2   Pre-processing Protocols

The PET analysis should be carried out in the template space, i.e., a brain atlas, to maximize the voxel correspondence across the subjects. All raw PET data in ADNI database have been spatially normalized with an isotropic voxel size of $1.5\,mm^3$ and a full width at half maximum resolution of $8\,mm$ [16]. As shown in Fig. 3.2-(1), the PET volumes were aligned to the MRI volumes of the same subject using FSL Linear Registration Tool (FLIRT) [17]. Then the MRI volumes were non-linearly registered to the ICBM_152 template [29] using the Image Registration Toolkit (IRTK) [31], as shown in Fig. 3.2-(2). IRTK works in a coarse-to-fine way with isotropic control point spaces downsizing from 12 to $1.5\,mm$ in 4 octaves. Lastly, the resultant coefficients of MRI registration were applied to the PET volume to map it to the same template, as shown in Fig. 3.2-(3). The ICBM_152 template is illustrated in Fig. 3.3, where different brain regions are displayed in different colors.

3.3   ADNI MRI-PET Subset

3.3.1   Screening Criteria

The MRI-PET subset is the intersection of the MRI subset and the PET subset and contains subjects with both MRI and PET data. A total of 352 subjects fulfill this criteria after the screening procedure in both subsets. However, 21 subjects were rejected during pre-processing, since there were errors reported by the registration algorithm when calculating the inverse registration for the MRI label maps. Therefore, the MRI-PET subset contains 331 subjects in total, including 85 AD, 169 MCI (102 ncMCI and 67 cMCI) and 77 NC subjects.

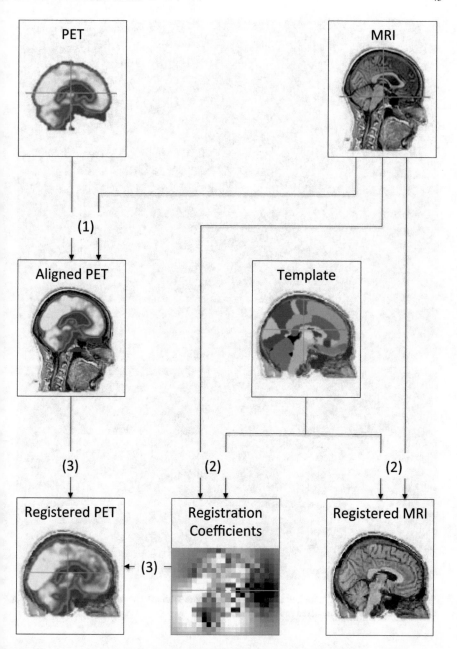

Fig. 3.2 The pre-processing pipeline for ADNI PET subset: (*1*) align PET to MRI; (*2*) non-linearly register MRI to a template; (*3*) apply registration coefficients onto PET. Figure reproduced with permission from [24]

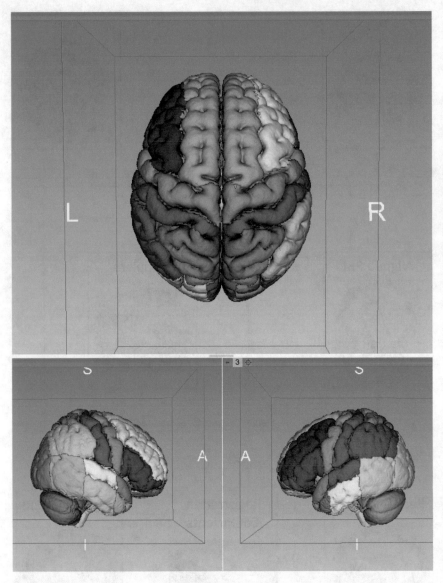

Fig. 3.3 The ICBM_152 template in the MNI coordinate system. The template is labeled with a total of 83 brain regions as listed in Appendix B

3.3.2 Pre-processing Protocols

The pre-processing of MRI-PET has two steps. The first step is to map the PET to MRI, same as the PET pre-processing described in Sect. 3.2, then calculate the inverse registration coefficients. The MRI label maps were than registered back to the original PET space. The purpose of inverse registration is to superimpose the label maps on PET, and segment it into different brain regions as in MRI volumes. Therefore, the anatomical information in MRI and the functional information in PET can be jointly analyzed.

3.4 ADNI dMRI Subset

3.4.1 Screening Criteria

The dMRI subset contained all of the 233 subjects recruited from the ADNI baseline cohort with ages ranging between 48 and 90. Each subject had a 3D axial brain dMRI scan [15]. The subjects were also labeled with diagnosis results, including 61 NC subjects, 123 MCI subjects and 49 AD subjects.

3.4.2 Pre-processing Protocols

We applied the the FSL Brain Extraction Tool (BET) to extract the brains from the dMRI data [33], and then modeled the diffusion tensors for each masked dMRI brain using the least square approach [18]. Whole-brain tractography was reconstructed based on the estimated tensors using a deterministic fiber-tracking algorithm [3]. The hyper-parameters used in the pre-processing pipeline are consistent across different subjects. Table 3.1 shows the parameters in constructing the tractography.

The Automated Anatomical Labeling (AAL) atlas was used to label the brain with predefined ROIs by registering the dMRI baseline template to the subject's baseline

Table 3.1 The parameters used in fiber tracking for the ADNI dMRI subset. Table reproduced with permission from [20]

Parameter	Value
Fractional Anisotropy (FA) Threshold	0.12
Angular Threshold	60
Step Size	0.68
Seed Number	10000

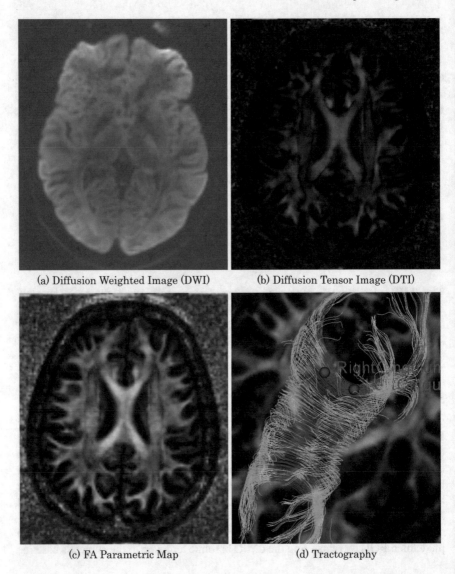

(a) Diffusion Weighted Image (DWI)   (b) Diffusion Tensor Image (DTI)

(c) FA Parametric Map   (d) Tractography

Fig. 3.4 The examples of (**a**) DWI, (**b**) DTI, (**c**) fractional anisotropy map and (**d**) tractography of the same subject

volume [36]. Measurements of the tracts, including the number, length and volume, were calculated to represent the regional tract density within the ROIs. Measurements of the tensors, including mean FA and ADC of the ROI, were calculated the average water diffusivity within the ROIs. Figure 3.4 illustrates an example of the dMRI data and the outputs of each step in the pre-processing pipeline. The reconstructed fibers were further filtered by the ROIs pair-wisely. The number of reconstructed fibers

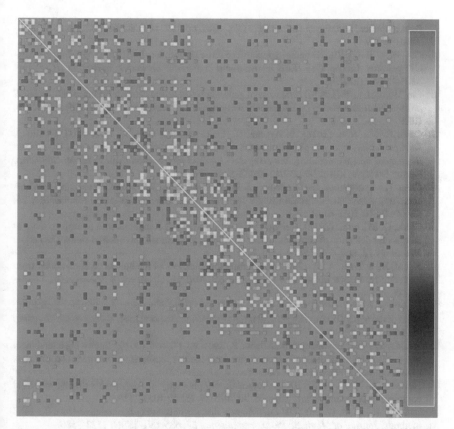

Fig. 3.5 An example of the symmetric brain inter-region matrix color map with AAL ROIs. Each element in matrix is the number of tracts filtered by a pair of ROIs. Figure reproduced with permission from [20]

between each pair of AAL grey matter ROIs is stored in a symmetric matrix. An example of the reconstructed matrix is shown in Fig. 3.5.

References

1. Ashburner, J., & Friston, J. K. (2000). Voxel-based morphometry - The methods. *NeuroImage*, *11*, 805–821.
2. Awate, S. P., Yushkevich, P. A., Song, Z., Licht, D. J., & Gee, J. C. (2010). Cerebral cortical folding analysis with multivariate modeling and testing: Studies on gender differences and neonatal development. *NeuroImage*, *53*(2), 450–459. ISSN:1053- 8119.
3. Basser, P., Pajevic, S., Pierpaoli, C., et al. (2000). In vivo fiber tractography using DTMRI data. *Magnetic Resonance Medicine*, *44*, 625–632.
4. Batty, S., Clark, J., Fryer, T., & Gao, X. (2008). Prototype system for semantic retrieval of neurological PET images English. In X. Gao, H. Müller, M. Loomes, R. Comley, & S. Luo

(Eds.), *Medical imaging and informatics* (Vol. 4987, pp. 179–188). Berlin: Springer. ISBN:978-3-540-79489-9.

5. Cai, W., et al. (2010). 3D neurological image retrieval with localized pathology-centric CMR-Glc patterns. In *The 17th IEEE international conference on image processing (ICIP)* (pp. 3201–3204). IEEE.

6. Cai, W., et al. (2014). A 3D difference of gaussian based lesion detector for brain PET. In *IEEE international symposium on biomedical imaging: From nano to macro (ISBI)* (pp. 677–680). IEEE.

7. Cash, D. M., et al. (2012). In Ayache, N., Delingette, H., Golland, P., & Mori, K. (Eds.), *Cortical folding analysis on patients with Alzheimer's disease and mild cognitive impairment in medical image computing and computer- assisted intervention* (Vol. 7512, pp. 289–296). Berlin: Springer. ISBN: 978-3-642-33453-5.

8. Chen, K., Ayutyanont, N., Langbaum, J. B., Fleisher, A. S., Reschke, C., et al. (2011). Characterizing Alzheimer's disease using a hypometabolic convergence index. *NeuroImage, 56,* 52–60. ISSN: 1053-8119.

9. Clark, C. M., et al. (2012). Cerebral PET with florbetapir compared with neuropathology at autopsy for detection of neuritic amyloid- Œ ≤ plaques: A prospective cohort study. *The Lancet Neurology 11,* 669–678. ISSN: 1474-4422.

10. Fischl, B., & Dale, A. M. (2000). Measuring the thickness of the human cerebral cortex from magnetic resonance images. *Proceedings of the National Academy of Sciences, 97,* 11050–11055.

11. Hammers, A., Allom, R., Koepp, M., et al. (2003). Three-dimensional maximum probability Atlas of the human brain, with particular reference to the temporal lobe. *Human Brain Mapping, 19,* 224–247.

12. Heckemann, R., Keihaninejad, S., Aljabar, P., Rueckert, D., et al. (2010). Improving intersubject image registration using tissue-class information benefits robustness and accuracy of multi-atlas based anatomical segmentation. *NeuroImage, 51,* 221–227.

13. Heckemann, R. A., et al. (2011). Automatic morphometry in Alzheimer's disease and mild cognitive impairment. *NeuroImage 56,* 2024–2037. ISSN: 1053-8119.

14. Jack, C. R., et al. (2008). The Alzheimer's disease neuroimaging initiative (ADNI): MRI methods. *Journal of Magnetic Resonance Imaging 27,* 685–691. ISSN: 1522-2586.

15. Jack, C., Bernstein, M., et al. (2010). Update on the mgnetic resonance imaging core of the Alzheimer's disease neuroimaging initiative. *Alzheimer's & Dementia, 6,* 212–220.

16. Jagust, W. J., et al. (2010). The Alzheimer's disease neuroimaging initiative positron emission tomography core. *Alzheimer's & Dementia 6,* 221–229. ISSN:1552-5260.

17. Jenkinson, M., Bannister, P., Brady, M., & Smith, S. (2002). improved optimization for the robust and accurate linear registration and motion correction of brain images. *NeuroImage, 17,* 825–841. ISSN: 1053-8119.

18. Koay, C., Chang, L., Carew, J., et al. (2006). A unifying theoretical and algorithmic framework for least squares methods of estimation in diffusion tensor imaging. *Journal of Magnetic Resonance, 182,* 115–125.

19. Landau, S. M., et al. (2013). comparing positron emission tomography imaging and cerebrospinal fluid measurements of beta-amyloid. *Annals of Neurology, 74,* 826–836. ISSN: 1531-8249.

20. Liu, S. Q., et al. (2015). Content-based retrieval of brain diffusion magnetic resonance image in multimodal retrieval in the medical domain 9059, Berlin: Springer.

21. Liu, S. Q., et al. (2015). Multi-modal neuroimaging feature learning for multi-class diagnosis of Alzheimer's disease. *IEEE Transactions on Biomedical Engineering, 62,* 1132–1140.

22. Liu, S., Cai, W., Wen, L., & Feng, D. (2013). Multiscale and multiorientation feature extraction with degenerative patterns for 3D neuroimaging retrieval. In *The 19th IEEE international conference on image processing (ICIP)* pp. (1249–1252). IEEE.

23. Liu, S., Cai, W., Wen, L., & Feng, D. (2013). Neuroimaging biomarker based prediction of Alzheimer's disease severity with optimized graph construction. In *IEEE international symposium on biomedical imaging: From nano to macro (ISBI)* (pp. 1324–1327). IEEE.

24. Liu, S., Liu, S. Q., Pujol, S., Kikinis, R., & Cai, W. (2014). Propagation graph fusion for multi-modal medical content-based retrieval. In *The 13th annual international conference on control, automation, robotics and vision (ICARCV)* (pp. 849–854). IEEE.
25. Liu, S., et al. (2013). A Supervised multiview spectral embedding method for neuroimaging classification. In *The 20th IEEE international conference on image processing (ICIP)* (pp. 601–65). IEEE.
26. Liu, S., et al. (2016). Cross-view neuroimage pattern analysis for Alzheimer's disease staging. *Frontiers in Aging Neuroscience*.
27. Liu, S., et al. (2013). Localized sparse code gradient in Alzheimer's disease staging. In *The 35th annual international conference of the IEEE engineering in medicine and biology society (EMBC)* (pp. 5398–5401). IEEE.
28. Liu, S., et al. (2011). Localized functional neuroimaging retrieval using 3D discrete curvelet transform. In *IEEE international symposium on biomedical imaging: From nano to macro (ISBI)* (pp. 1877–1880). IEEE.
29. Mazziotta, J., et al. (2001). A Probabilistic Atlas and reference system for the human brain: International consortium for brain mapping (ICBM). *Philosophical Transactions of the Royal Society of London. Series B: Biological Sciences 356*, 1293–1322.
30. Minoshima, S., Frey, K. A., Koeppe, R. A., Foster, N. L., & Kuhl, D. E. (1995). A diagnostic approach in Alzheimer's Disease using three-dimensional stereotactic surface projections of Fluorine-18-FDG PET. *Journal of Nuclear Medicine, 36*, 1238–1248.
31. Rueckert, D., Sonoda, L., Hayes, C., et al. (1999). Non-rigid registration using free-form deformations: Applications to breast MR images. *IEEE Transactions on Medical Imaging, 18*, 712–721.
32. Schaer, M., et al. (2008). A surface-based approach to quantify local cortical gyrification. *IEEE Transactions on Medical Imaging, 27*, 161–170.
33. Smith, S. (2002). Fast robust automated brain extraction. *Human Brain Mapping, 17*, 143–155.
34. Sokoloff, L., Reivich, M., Kennedy, C., Des-Rosiers, M., et al. (1977). The [14C]Deoxy-Glucose method for the measurement of local cerebral glucose utilization: Theory, procedure and normal values in the consicious and anesthetized albino rat. *Journal of Neurochemistry, 28*, 897–916.
35. Suk, H.-I., Lee, S., & Shen, D. (2013). Latent feature representation with stacked auto-encoder for AD/MCI diagnosis. *Brain Structure and Function, 220*, 841–959.
36. Tzourio-Mazoyer, N., Landeau, B., Papathanassiou, D., Crivello, F., et al. (2002). Automated anatomical labelling of activations in SPM using a macroscopy anatomical pacellation of the MNI MRI single-subject brain. *NeuroImage, 15*, 273–289.

Chapter 4
Encoding the Neurodegenerative Features

Various feature descriptors have been proposed to model the disease pathologies. Chapter 2, Sect. 2.2 reviews the common morphological, functional and learning-based features. These features are capable of capturing many important changes in the brain, such as brain atrophy and hypo-metabolism, but they are constrained to detect other subtle and complicated changes, such as shape of the cortical regions or contrast between the lesions and normal tissues. In addition, it is also very challenging to fuse the multimodal features. We therefore proposed a set of hand-engineered features and learning-based features to supplement them.[1]

This chapter is organized as follows. Section 4.1 describes the methods for modeling the brain morphological features, and shows the classification results of the Alzheimer's Disease Neuroimaging Initiative (ADNI) MRI subset. Section 4.2 gives the details for modeling the functional features, which are validated on the ADNI PET subset. The learning-based methods are discussed in Sect. 4.3, and the extracted features are validated using the ADNI MRI-PET subset.

4.1 Encoding the Morphological Features

Although many brain morphological feature descriptors have been widely used in previous studies, such as volume [14, 15] and contour/texture [4, 9–13, 24] of the cortical regions, it is impossible for them to capture all of the anatomical differences between subjects. For example, the hippocampus is a small organ located in the medial temporal lobe and associated with memory. Its size is an important biomarker in mild cognitive impairment (MCI) and Alzheimer's disease (AD). Table 4.1 shows the left hippocampus and the feature values of four examples from different groups of subjects, including normal control (NC), MCI non-converter (ncMCI), MCI converter

---

[1]Some content of this chapter has been reproduced with permission from [5, 8, 15, 17].

© Springer Nature Singapore Pte Ltd. 2017
S. Liu, *Multimodal Neuroimaging Computing for the Characterization of Neurodegenerative Disorders*, Springer Theses, DOI 10.1007/978-981-10-3533-3_4

Table 4.1 Left hippocampus and their morphological feature values of four subjects, diagnosed with NC, ncMCI, cMCI and AD, respectively. Table reproduced with permission from [17]

	NC	ncMCI	cMCI	AD
E.g., left hippocampus				
Volume (%)	1.57	1.57	1.57	1.57
Solidity (%)	85.78	73.05	67.35	79.32
Convexity (%)	84.31	84.34	86.06	85.77

(cMCI) and AD. All of the four subjects have identical feature value in grey matter volume (GMV), although the AD subject presented clear atrophy. This demonstrates that, in this specific case, GMV is not able to distinguish these 4 subjects. However, two other morphological features, namely the convexity and solidity, which will be discussed later in this section, are effective to detect the subtle atrophic changes.

4.1.1 Convex-Based Morphological Feature Descriptors

As shown in Table 4.1, the convexity and solidity features could provide the complementary information to the conventional GMV in describing the brain region atrophy. The design of these two feature descriptors is inspired by local gyrification index (LGI), which is used to capture the focal cortical folding in each vertex on the surface. However, the computing of LGI is very time-consuming and does not support the ROI-level analysis. Both convexity and solidity require constructing the convex hull of the regions of interest (ROIs). Figure 4.1 shows three examples of the reconstructed brain ROI surfaces and their convex hulls.

Convexity Ratio (CNV) is defined as the area of the convex hull surface divided by that of the ROI surface [15], as in Eq. 4.1, where S is the surface area. It is used to quantify the cortical foldings in cortical ROIs, similar to LGI. A normal brain with extensive cortical folding usually has a greater CNV, whereas an atrophic brain with less foldings may have smaller CNV.

$$CNV = \frac{S_{Convex\_Hull}}{S_{ROI}} \tag{4.1}$$

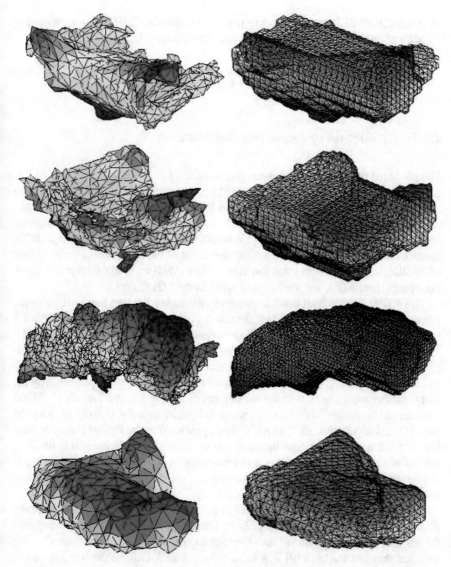

Fig. 4.1 The surface model and convex hull of three cortical regions. *Left column* the reconstructed surface models; *right column* the corresponding 3D convex hulls of the surface models in the *left column*

Solidity Ratio (SLD), is defined as the ratio of volume of the ROI to that of the convex hull, as in Eq. 4.2 where V is the volume. It is used to describe the fullness of the ROI in its convex hull, either convex or concave. The degenerated brains are usually in a wrinkled shape compared to normal healthy brains, thus present a lower

SLD value. Both SLD and CNV can be used to enhance the GMV features, especially when there are large inter-subject brain volume variations [15].

$$SLD = \frac{V_{ROI}}{V_{Convex\_Hull}} \tag{4.2}$$

4.1.2 Evaluation of Convex-Based Features

The proposed features were validated on the ADNI MRI subset. The details of this dataset can be found in Chap. 3, Sect. 3.1. The proposed morphological features were compared with the widely used GMV and LGI features. LGI features are usually computed in each hemisphere. To match the other ROI-level feature descriptors, here we customized the LGI descriptor to estimate the gyrification index in ROIs instead, i.e., calculating the mean gyrification index of the vertices on the our surface of the ROI, instead of the gyrification index of individual vertices on the pial surface. For non-cortical ROIs, we used surface areas as the LGI features.

The ADNI MRI subset used in this analysis highly favored the MCI subjects, which accounted for more than half the entire population. We therefore designed 3 cascading classifiers instead of one single classifier to reduce the data bias onto classification. The first and second classifiers were both binary support vector machines (SVMs). The first cascading classifier was used to distinguish NC subjects from non-NC subjects, including both AD and MCI patients. The NC subjects predicted by the first classifier were kept out of the subsequent classification, and the other subjects were sent to the second classifier. The second classifier aimed to classify the subjects into AD or non-AD patients. The AD patients predicted by the second classifier were kept out of the subsequent classification, and the other subjects were sent to the third cascading classifier, a multi-class SVM that classified AD, MCI and NC subjects in one setting. All three SVMs were kernelized with Radial-Basis-Function (RBF) kernel and implemented using LIBSVM library [6].

The 5-fold cross-validation was used in performance evaluation, i.e., we equally divided the datasets into five subsets. In each validation trial, four subsets were used for training the model, whereas the other subset was used for testing the model. We repeated this process until all five subsets were tested. During the training phase, the three SVMs were trained simultaneously to optimize the hyper-parameters using the random search optimization algorithm [3]. We used 6 metrics in this study to evaluate the classification performance, including precision (for AD, MCI and NC, respectively), accuracy, specificity and sensitivity of all groups combined together. Notice specificity and sensitivity can only be computed on two classes, we therefore consider NC as the negative class, and both MCI and AD as the positive class. We also reported the standard deviations of these metrics in cross-validation.

The classification performances of the individual morphological features are summarized in Tables 4.2 and 4.3. It was evident that no single morphological feature could completely outperform the others. The best result of each performance met-

Table 4.2 The classification precision (%) of NC, MCI and AD subjects in the ADNI MRI subset using the MRI features. Table reproduced with permission from [17]

Feature		NC	MCI	AD
MRI	GMV	43.08 ± 11.48	**60.23 ± 8.30**	**67.64 ± 22.93**
	LGI	31.27 ± 6.38	52.34 ± 12.98	44.04 ± 12.75
	CNV	**43.91 ± 4.28**	57.57 ± 4.25	50.63 ± 12.94
	SLD	39.86 ± 8.60	55.64 ± 3.56	47.09 ± 5.66

Table 4.3 The classification accuracy, specificity and sensitivity (%) of subjects in the ADNI MRI subset using the MRI features. Table reproduced with permission from [17]

Feature		Accuracy	Specificity	Sensitivity
MRI	GMV	**55.58 ± 5.49**	**48.33 ± 10.26**	78.76 ± 11.68
	LGI	44.42 ± 9.78	35.17 ± 7.87	76.35 ± 4.37
	CNV	52.27 ± 5.92	47.58 ± 17.96	**81.45 ± 7.38**
	SLD	49.24 ± 1.99	48.17 ± 16.29	78.37 ± 4.50

ric is highlighted in bold-face. In general, GMV tended to have the best overall performance, especially on, MCI precision, AD precision and the overall accuracy and specificity. The second best feature is CNV, which had highest NC precision and sensitivity. However, SLD has a competing performance with GMV in overall specificity and sensitivity.

4.2 Encoding the Functional Features

Positron emission tomography (PET) provides important insights into neuroscience and has an essential role in the computer-aided-diagnosis (CAD) of neurological disorders and neuro-oncology applications. FDG-PET enables the detection of the brain lesions with metabolism anomalies, even before these lesions can be perceived in MRI. Current PET-based functional feature descriptors, such as t-maps [4], mean index [1] and z-scores [18], may capture the location of the lesions and/or the degree of their abnormality, but they require a bunch of normal subjects to define the normal range of values and rely heavily on image registration techniques. In addition, many important properties of the lesions, such as contrast of lesional and non-lesion tissues are ignored. To capture such important information, we proposed an automatic 3D lesion detector and three lesion-centric feature descriptors for FDG-PET data.

4.2.1 Neurodegenerative Lesion Detection

The tissues in a brain ROI can be classified into lesional and non-lesion tissues. Ideally, a lesion detector should effectively detect the salient regions, either with hyper- or hypo-metabolism, in a form that is invariant to the changes in image scales. To ensure the scale-invariance, we represented an image I as a function of location x in an image scale-space $I(x, \sigma)$, as in Eq. 4.3:

$$I(x, \sigma) = I(x, \sigma_0) * G(x, \sigma - \sigma_0) \tag{4.3}$$

where $G(x, \sigma - \sigma_0)$ is a Gaussian filter with a variance of σ^2, and σ_0^2 is the variance of the initial selected Gaussian filter. The Gaussian scale-space $I(x, \sigma)$ of an image is defined as the convolution product of the image and the filter.

We assumed that the metabolism rates of the lesion tissues are distributed with smooth densities, and reformulated salience detection in the scale-space as detection of the maximum changes in the voxel values with respect to the changes in voxel locations and scales. In this study, we used the Difference-of-Gaussian (DoG) operator to infer the lesion candidates, which have such extrema changes, as in Eq. 4.4:

$$(x, \sigma) = local \arg \max_{x,\sigma} |\frac{dI(x, \sigma)}{d\sigma}| \tag{4.4}$$

where $x = (x_1, x_2, x_3)$ is the 3D coordinate of a local extrema detected by the isotropic 3D Gaussian filter with a standard deviation of σ, and each pair of (x, σ) points to a lesion candidate. The 3D DoG operator has been previously used in MRI morphometry to detect the key points [23], but we found it was very difficult to use these salient points as features since they vary in geometry and appearance from one subject to another. However, the PET signals are more robust to DoG operators, since lesions on PET are generally in spherical shapes, which are less sensitive to local structural changes.

In this analysis, we divided each octave of the scale space into 3 intervals, i.e., scale factor $k = \sqrt{2}$. The initial scale σ_0 was selected as 1 voxel, which corresponds to a full width at half magnitude (FWHM) of $4 \times \sqrt{2} \times \log(2)$ voxels, and the scale increase by a factor of k, as $\sigma_i = \sigma_0 \times k^i$. Totally nine DoG volumes were produced over two octaves, corresponding to the Gaussian filters with σs of five to 19 voxels. Take a $128 \times 128 \times 128$ PET scan with voxel size $2\,mm \times 2\,mm \times 2\,mm$ for instance; the proposed algorithm is able to detect lesions with a diameter from 9.42 to 37.68 mm.

We designed a set of screening criteria to filter out the false positive lesion candidates from the detected lesion set S. The lesion candidates with a high curvature ratio γ were first excluded ($\gamma = 5$). The curvature ratio was calculated based on a small patch of $3 \times 3 \times 3$ voxels around the center x of each detected lesion candidates in the 3D space. Three Eigen vectors pointing to three principle directions were computed from this patch, and then the curvature ratio for each center was calculated as the ratio of the largest Eigen value to the smallest Eigen value. The salient lesion candidates with a curvature ratio larger than γ were then filter out. Secondly, the lesion candidates with minimum values in their local neighborhoods were excluded, not like in other studies using both maxima and minima, since the focus here was the neurodegenerative lesions with lower uptake values only. Finally, we noticed there was a harmonic effect when using convolution with a set of Gaussian filters. Therefore, if a lesion was detected as a local minima for multiple times, we choose the one with the smallest value across the scale-space. Figure 4.2 illustrates the detected lesional tissues on PET using the proposed DoG operator. Figure 4.2a shows the lesion patterns of three randomly selected AD patients, Fig. 4.2b shows three MCI patients, and Fig. 4.2c shows three NC subjects.

4.2.2 DoG-Based Functional Feature Descriptors

Three feature descriptors were proposed to quantitatively describe the detected lesions. All of the three features were designed to be invariant to the actual size of the PET scans, and robust to the local anatomical changes.

Difference-of-Gaussian Mean, DoG-M, is defined in Eq. 4.5 where M is the mean value. DoG-M quantifies the degeneration levels of the lesions estimated by the DoG detector. It is defined as the mean metabolism rate of the lesional tissues [5]. DoG-M is different from the conventional mean index [1] in that it focuses on the metabolism level of the lesional area only. To eliminate the bias of global intensity variation across different scans, the DoG-M values were further normalized by the mean metabolism rate of the cerebellum.

$$DoG_M = \frac{M_{Lesion}}{M_{ROI}} \tag{4.5}$$

Difference-of-Gaussian Contrast, DoG-C, is defined in Eq. 4.6 where M is the mean value, VA is the variance, and V is the volume. DoG-C quantifies the contrast between the lesional and non-lesion tissues. The metabolism rates may have a wide range of values across different ROIs. DoG-C compensates this negative impact by calculating the contrast instead of the actual metabolism rate of the lesional tissues. It was previously known as the lesion contrast index [5], which was defined as the ratio of the mean metabolism rate in lesions to that in non-lesion tissues. DoG-C was further corrected using the variances of both lesional and non-lesion tissues in the ROIs.

$$DoG_C = \log \frac{M_{Lesion} - M_{Non\_Lesion}}{\sqrt{\frac{VA_{Lesion}}{V_{Lesion}} + \frac{VA_{Non\_Lesion}}{V_{Non\_Lesion}}}} \tag{4.6}$$

Difference-of-Gaussian Z-score, DoG-Z, which is similar to the conventional Z-score [18], quantifies the proportion of the abnormal voxels in the ROIs. AD patients at their mild and late stage usually have higher z-score DoG-Z values than patients at

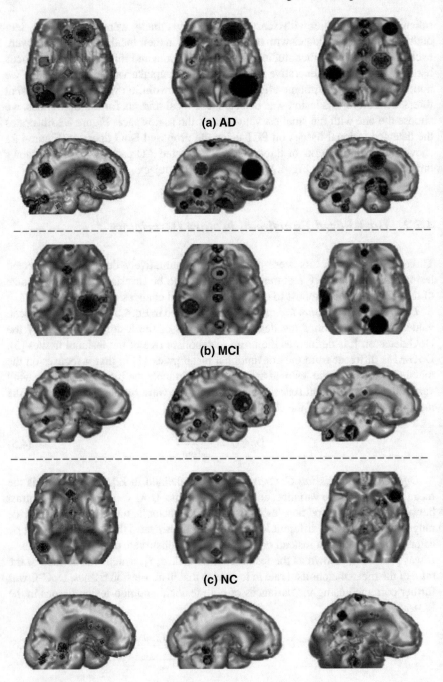

Fig. 4.2 Examples of the detected lesional tissues of subjects in three population groups, including **a** AD, **b** MCI and **c** NC. Figure reproduced with permission from [5]

their early stage. The drawback of the conventional z-score is that it requires repeated image registration between a bunch of normal subjects and a brain template to derive the z-score on each voxel, thereby bringing in many registration errors. The DoG operator, on the other hand, estimates the hypo-metabolism lesions without involving image registration [5], and DoG-Z is a more robust indicator to approximate the volume of the lesional tissues.

$$DoG_Z = \frac{V_{Lesion}}{V_{ROI}} \qquad (4.7)$$

4.2.3 Evaluation of the DoG-Based Features

The proposed DoG features were validated on the ADNI PET subset, as described in Chap. 3, Sect. 3.2. The proposed functional features were compared with the mean index (M-IDX) [1] and fuzzy index (F-IDX) [15] feature descriptors. M-IDX, defined as the mean metabolism rate of the ROIs, is a simple and powerful feature in modeling brain metabolism abnormalities and commonly used to support the diagnosis of AD and MCI. It is particularly sensitive to hypo-metabolism in MCI and outperformed many complex feature descriptors [16], such as 3D Gabor Filters [9], and Gray Level Co-occurrence Matrix [4, 11], and Discrete Curvelet [12, 13]. To correct the artifacts in intensity variations raised at image acquisition or parameter estimation, we further normalized the M-IDX values using the mean metabolism rate in cerebellum.

F-IDX evaluates the heterogeneity, or 'fuzziness', of the metabolism rates in ROIs, which is defined as the standard deviation divided by the mean metabolism rate in the ROI. It aims to address the partial volume effect in these ROIs that contain both normal and hypo-metabolic areas and have high heterogeneous metabolism activities, hence a higher F-IDX. Those ROIs with high homogeneous metabolism activities, either pathological or normal, are expected to have more consistent values, hence a smaller F-IDX.

Same design of the cascading classifiers and performance metrics as in Sect. 4.1.2 were used to evaluate the performance of the proposed functional features. Tables 4.4 and 4.5 show the classification results of the proposed DoG features compared to

Table 4.4 The classification precision (%) of NC, MCI and AD subjects in the ADNI PET subset using the PET features. Table reproduced with permission from [17]

Feature		NC	MCI	AD
PET	M-IDX	45.46±7.96	60.88±3.57	**62.19±13.46**
	F-IDX	**53.30±10.55**	**64.06±3.69**	49.47±6.32
	DoG-M	49.02±14.52	63.20±7.12	55.39±11.66
	DoG-C	44.84±8.92	57.56±5.33	55.83±7.00
	DoG-Z	33.84±2.41	51.95±3.62	51.91±8.26

Table 4.5 The classification accuracy, specificity and sensitivity (%) of subjects in the ADNI PET subset using the PET features. Table reproduced with permission from [17]

Feature		Accuracy	Specificity	Sensitivity
PET	M-IDX	56.20 ± 3.06	51.83 ± 10.61	81.11 ± 4.27
	F-IDX	**56.49 ± 3.89**	**63.67 ± 8.52**	81.90 ± 7.75
	DoG-M	56.19 ± 4.68	50.58 ± 4.20	**82.27 ± 7.99**
	DoG-C	52.88 ± 4.56	52.33 ± 19.49	79.47 ± 10.27
	DoG-Z	46.54 ± 2.89	42.92 ± 8.83	74.82 ± 3.06

other popular features. F-IDX was the best feature descriptor with the highest NC precision (53.30%), MCI precision (64.06%), overall accuracy (56.49%) and specificity (63.67%). M-IDX achieved the highest precision in AD classification (62.19%), and DoG-M had the best performance in overall sensitivity (82.27%).

4.3 Encoding the Multimodal Features

Many machine learning methods have been proposed to aid the characterization of neurodegenerative disorders using multimodal data, as elaborated in Chap. 2, Sect. 2.2.3. A number of studies attempted to select the most representative features in each feature space using supervised methods, such as elastic net and t-test, and then fuse the features to form a new feature space [7, 14, 21]. The classifiers will then be trained with the features in the new feature space. Such approaches may have two limitations, (1) result in large repetition and inconsistency across different feature spaces using the linear feature selection methods, and (2) lead to information lose due to the two training goals in feature selection and data fusion. To overcome these limitations in previous approaches, we therefore designed a new deep-learning framework which can efficiently encode the multimodal features and effectively characterize the neurological disorders at the same time.

4.3.1 Multimodal Deep Learning Framework

Stacked auto-encoder [19, 20], which is an advanced unsupervised neural network for decomposing the original data into building blocks, was used to learn the high-level features. As shown in Fig. 4.3, each auto-encoder encodes an input vector x into a hidden representation y with a linear mapping followed by a non-linear sigmoidal distortion:

Fig. 4.3 Illustration of the single-modal and multimodal deep learning architectures of the proposed framework. Figure reproduced with permission from [8]

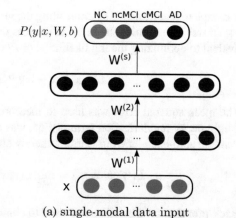

(a) single-modal data input

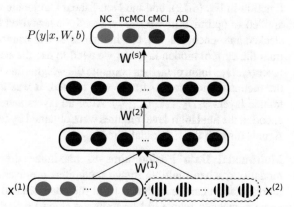

(b) multi-modal data input with one modality
hidden from the other

$$y = \sigma(Wx + b) \tag{4.8}$$

$$\sigma(*) = \frac{1}{1 + e^{-*}} \tag{4.9}$$

where σ is the sigmoid function, W and b are weights and bias terms in the hidden layer, and y is the encoding of the original input x. Ideally, we can losslessly reconstruct input vector x from the encoding y and the weights and bias in the hidden layer:

$$x^* = \sigma(W'y + b') \tag{4.10}$$

where σ is another sigmoid function, W' is the decoding weights, and b' is the decoding bias terms. The dimensionality of y is determined by the number of the hidden neurons at the hidden layer, and it is effortless to obtain a condensed or over-complete feature space by controlling the number of nodes in the hidden layer. Our goal was to minimize the information loss in the reconstruction of x, which is equivalent to optimizing the log likelihood of $P(x^*|x)$:

$$L(x, x^*) \propto -\log P(x^*|x) \tag{4.11}$$

The mean squared error was used to measure the reconstruction loss $L(x, x^*)$. A weight decay regularization term, $\|W\|_2^2$, was added to the objective function to prevent the auto-encoder from learning merely an identity function:

$$L(W, b, x, x^*) = \min_{W,b} L(x, x^*) + \lambda \|W\|_2^2 \tag{4.12}$$

Back-propagation algorithm was used to compute the gradients of the objective function in Eq. (4.12), and the Non-Linear Conjugate Gradient algorithm [19] was applied to optimize it. We used a greedy layer-wised training strategy to train the stacked auto-encoders, i.e., one hidden layer was trained each time. For example, to train the current hidden layer l_n, we need to use the outputs of the previous hidden layer (l_{n-1}) as input data, then compute the weights and bias terms in l_n that minimize the reconstruction cost. After l_n was trained, it was then stacked on the previous trained layers $(l_1, l_2, \ldots, l_{n-1}, l_n)$. After all layers were trained in the stacked auto-encoder, the final high-level features were obtained by feed-forwarding the activation signals through the sigmoidal filters.

Multimodal Data Fusion aims to maximize the synergy between different modalities, when multiple image modalities are available in the analysis. Shared representation can be obtained from the concatenated MRI and PET inputs using an auto-encoder, whole hidden units are able to model the correlation between the dimensions of both MRI and PET. Such correlation can also be computed with linear algorithms like Principal Component Analysis (PCA), whole Eigen vectors form the transformations that maximize the variance in the MRI and PET data. However, the correlations of MRI and PET are highly non-linear, and such simple concatenation strategy always results in the hidden neurons activated by one single modality. To overcome these drawbacks of concatenation strategy and enhance the robustness of the proposed stacked auto-encoder, we employed a 'zero masking' strategy [19], that only uses the corrupted inputs of individual modalities to train the deep neural network. In other words, we randomly 'corrupted' the training samples of one modality by setting their values to zero, and then kept the other modality untouched when training the stacked auto-encoder. The first hidden layer of the stacked auto-encoder reconstructed the original inputs from the corrupted inputs. Then the outputs of the first layer propagated to higher layers in the same network. Each higher layer was then trained progressively to learn representations to reconstruct the original inputs from the noisy representations of the corrupted inputs. With this strategy, the hidden neurons in our stacked auto-encoder could be activated by both modalities, such able to infer their nonlinear correlations.

Softmax Logistic Regression is a supervised algorithm designed for multi-class classification problems. Although the features extracted by the stacked auto-encoder can be used to train conventional classifiers, such as SVM, softmax logistic regression can be seamlessly combined with stacked auto-encoders, thus enables us to optimize the parameters in itself and the connected stacked auto-encoder via fine-tuning. In this analysis, the softmax layer was connected to the last hidden layer of the stacked auto-encoder. Features extracted by the stacked auto-encoder were used as the inputs in softmax regression [2], then the activation signals were propagated through the entire neural network and optimized all the parameters according to the classification loss as a supervised neural network, as shown in Fig. 4.3a. When training the supervised network with the multimodal inputs, a small proportion of the single modal inputs were dropped out using the zero masking strategy, then the hidden neurons were trained to make predictions, as shown in Fig. 4.3b.

The activation function used in the softmax layer was different from the stacked auto-encoder's activation function, as defined in Eq. 4.13:

$$P(Y = i|x) = \frac{e^{W_i^{(s)}a + b_i^{(s)}}}{\sum_i e^{W_i^{(s)}a + b_i^{(s)}}} \qquad (4.13)$$

where Y is the predicted label, a is the feature representation obtained from the last hidden layer of the stacked auto-encoder, and $W_i^{(s)}$ and $b_i^{(s)}$ are the weights and bias terms for the i-th possible prediction.

In this analysis, we modeled the diagnosis of AD as a multi-class classification problem including four pre-defined labels: NC, cMCI, ncMCI and AD. For example $P(Y = \text{'AD'} |x^{(l)})$, is the probability of a patient being diagnosed as AD. The label with the highest probability was the predicted label. Softmax has the same objective function as auto-encoder, as defined in Eq. 4.14:

$$L(W, b, X, Y) = \min_{W,b} J(X, Y) + \lambda^{(s)} \| W^{(s)} \|_2^2 \qquad (4.14)$$

where W and b are the weights and bias terms of the hidden layers, including the pre-trained stacked auto-encoder and the softmax regression layer, $J(X, Y)$ is the logistic regression cost of the predicted label generated from X and the true label Y, $\lambda^{(s)}$ is the hyper-parameter that regulates weight decay on the softmax layer.

4.3.2 Evaluation of the Deep-Learning Features

To evaluate the proposed deep learning framework, the proposed stacked auto-encoder trained with zero masking method (SAE-ZERO) was compared to: (1) the SVM trained with MRI features only (SVM-MRI), (2) the SVM trained with PET features only (SVM-PET), (3) the stacked auto-encoder trained with MRI features (SAE-MRI), (4) the stacked auto-encoder trained with PET features only (SAE-

PET), (5) the multiple kernel SVM trained with multimodal features (MK-SVM), (6) the stacked auto-encoder trained with the concatenation strategy, and (7) the Suk's method that trains two stacked auto-encoders independently and then fuses their output features with a multiple kernel SVM [22]. The GMV and M-IDX were used as the multimodal inputs of MRI and PET. Ten-fold cross-validation paradigm was used in performance evaluation, with with 90% of the datasets for training and fine-tuning the model and the rest 10% for testing the model in each validation trial.

We set two hidden layers in all deep-learning methods, since our pilot experiments did not show any improvement with more hidden layers. We also enforced the hidden layers to have the same number of neurons, as suggested by Bengio et al. [2]. The this hyper-parameter, number of neurons, for individual models were searched by the approx search in log-domain approach [3] starting from 30 to 200.

Tables 4.6 and 4.7 summarize the performance of these methods. It was most challenging to classify ncMCI and cMCI subjects, since these two groups were highly similar in their pathologies. The best performance on these two groups are 52.76 and 53.17%, respectively. Multimodal methods (MK-SVM, SAE-CONCAT, SAE-MK-SVM and SAE-ZERO) achieved much better results than the single-modal methods

Table 4.6 The classification precision (%) of NC, MCI and AD subjects in the ADNI MRI-PET subset using the multimodal features. Table reproduced with permission from [8]

Methods	NC	ncMCI	cMCI	AD
SVM-MRI	49.74 ± 8.79	44.58 ± 14.91	46.45 ± 31.63	53.74 ± 10.20
SVM-PET	30.30 ± 20.15	36.90 ± 11.63	45.79 ± 27.08	50.30 ± 7.00
SAE-MRI	47.80 ± 17.97	40.39 ± 9.46	45.08 ± 24.95	56.33 ± 14.03
SAE-PET	41.79 ± 11.76	35.17 ± 10.10	41.06 ± 10.06	54.25 ± 11.79
MK-SVM	47.71 ± 12.73	**52.76 ± 19.33**	38.17 ± 31.94	53.81 ± 6.81
SAE-CONCAT	49.21 ± 14.74	43.54 ± 9.43	49.62 ± 9.66	56.35 ± 14.21
SAE-MK-SVM	53.86 ± 11.47	52.08 ± 18.65	**53.17 ± 26.63**	55.58 ± 13.06
SAE-ZERO	**59.07 ± 19.74**	52.21 ± 11.84	40.17 ± 14.42	**64.07 ± 15.24**

Table 4.7 The classification accuracy, specificity and sensitivity (%) of subjects in the ADNI MRI-PET subset using the multimodal features. Table reproduced with permission from [17]

Methods	Accuracy	Sensitivity	Specificity
SVM-MRI	47.74 ± 1.82	**66.43 ± 14.46**	78.78 ± 8.13
SVM-PET	42.60 ± 2.90	35.36 ± 23.00	79.95 ± 8.33
SAE-MRI	45.61 ± 8.31	48.04 ± 14.97	82.69 ± 7.88
SAE-PET	42.91 ± 6.63	43.04 ± 17.45	82.26 ± 5.36
MK-SVM	48.65 ± 4.29	61.07 ± 18.95	79.86 ± 6.43
SAE-CONCAT	48.96 ± 5.32	46.61 ± 22.04	84.63 ± 8.51
SAE-MK-SVM	51.39 ± 5.64	66.25 ± 18.34	82.66 ± 6.16
SAE-ZERO	**53.79 ± 4.76**	52.14 ± 11.81	**86.98 ± 9.62**

that using MRI or PET features alone. The only exception is SVM-MRI in sensitivity (66.43%), although SAE-MK-SVM had almost the same performance. In multimodal methods, deep learning methods (SAE-CONCAT, SAE-MK-SVM and SAE-ZERO) had the best classification performance in all patient groups, except the ncMCI group. SAE-ZERO had a competitive performance with MK-SVM in ncMCI, yet with a much smaller standard deviation. SAE-ZERO also outperformed the other methods in NC precision (59.04%), AD precision (64.07%), overall accuracy (53.79%) and overall specificity (86.98%), which were 5.2, 7.7, 2.4, and 2.3% higher than the second best performance.

4.4 Summary

Hand-Engineered Features The hand-engineered features, including both the MRI and PET features, were validated in the AD/MCI/NC classification experiments using the ADNI MRI and PET subsets. The PET features tended to have better performance than MRI features with higher overall specificity and precision in recognizing NC and MCI subjects. The only exception was DoG-Z, which had lower NC precision, MCI precision, accuracy and sensitivity compared to the MRI features. The sensitivity was remarkably higher than the specificity across all the features, with an average difference of 30.44%. The underlying reason for this observation is that both AD and MCI were considered as the positive class when computing the sensitivity and specificity. We argue that sensitivity is important in the diagnosis of AD, since a higher sensitivity means a higher chance to detect the positive class (AD and MCI), thus avoid treating the patients as normal subjects. It was very clear that no single feature descriptor could win all. F-IDX achieved the highest NC precision (53.30%), MCI precision (64.06%), overall accuracy (56.49%) and specificity (63.67%). GMV had the highest AD precision (67.64%), and DoG-M had the highest sensitivity (82.27%). This fact implied that different features had different strengths in the characterization of AD and MCI, and potentially there is room for further improvement in the performance when these features are combined.

Learning-based Features The learning-based features were validated in the AD/cMCI/ncMCI/NC classification experiments using the ADNI MRI-PET subset. The multimodal methods, such as MK-SVM, SAE-MK-SVM and SAE-ZERO, could benefit from the multimodal inputs, thus achieve better performance than methods trained with single modal inputs or concatenated inputs. The proposed SAE-ZERO also avoided training independent models on different modalities, therefore preserved the complementary information when learning the high-level features. When trained with the corrupted inputs, neurons in the stacked auto-encoder tent to be activated by both MRI and PET inputs. As shown in our most challenging experiments, i.e. detecting the ncMCI and cMCI subjects, the proposed SAE-ZERO method was superior to other methods in detecting the subtle differences between ncMCI and cMCI subjects, and also more robust to outliers in the training set.

Although other conventional classifiers, such as SVM, can also be used to classifier the high-level features, softmax regression is believed a better choice. The softmax layer on top of a stacked auto-encoder can seamlessly combine the supervised and unsupervised models, and enable the fine-tuning of the parameters. We argue that the fine-tuning made the key contribution to the performance. In our follow-up experiments, we transferred the features learnt from the fine-tuned neural network to other popular classifiers, such as SVM and KNN, all these classifiers showed highly consistent performance.

The methods with greater 'depth', a notion that describes the complexity of the algorithm, could capture the nonlinear relationship between multimodal inputs, thus improve the performance, as demonstrated by the comparison between SAE-MK-SVM and MK-SVM. Compared to traditional methods, the proposed deep learning framework was more powerful in extracting the high-level features in the multimodal data.

References

1. Batty, S., Clark, J., Fryer, T. & Gao, X (2008). Prototype System for Semantic Retrieval of Neurological PET Images English. In Gao, X., Mäller, H., Loomes, M., Comley, R. & Luo, S. (Eds.) Medical Imaging and Informatics, pp. 179–188, 4987. Springer: Berlin Heidelberg. ISBN: 978-3-540-79489-9.
2. Bengio, Y., Courville, A., & Vincent, P. (2013). Representation learning: A review and new perspectives. *IEEE Transactions on Pattern Analysis and Machine Intelligence, 35*, 1798–1828.
3. Bergstra, J., & Bengio, Y. (2012). Random search for hyper-parameter optimization. *Journal of Machine Learning Research, 13*, 281–305. issn: 1532-4435.
4. Cai, W. et al. (2010). 3D Neurological Image Retrieval with Localized Pathology-Centric CMRGlc Patterns in The 17th IEEE International Conference on Image Processing (ICIP), pp. 3201–3204. IEEE.
5. Cai, W. et al. (2014). A 3D Difference of Gaussian based Lesion Detector for Brain PET in IEEE International Symposium on Biomedical Imaging: From Nano to Macro (ISBI), pp. 677–680. IEEE.
6. Chang, C. C., & Lin, C. J. (2011). LIBSVM: A library for support vector machines. *ACM Transactions on Intelligent Systems Technology (ACM TIST), 2*, 27:1–27:27. issn: 2157-6904.
7. Che, H. et al. (2014). Co-Neighbor Multi-View Spectral Embedding for Medical Contentbased Retrieval in IEEE International Symposium on Biomedical Imaging: From Nano to Macro (ISBI), pp. 911–914. IEEE.
8. Liu, S. Q., et al. (2015). Multi-modal neuroimaging feature learning for multi-class diagnosis of alzheimers disease. *IEEE Transactions on Biomedical Engineering, 62*, 1132–1140.
9. Liu, S., Cai, W., Wen, L. & Feng, D. (2012). Multiscale and Multiorientation Feature Extraction with Degenerative Patterns for 3D Neuroimaging Retrieval in The 19th IEEE International Conference on Image Processing (ICIP), pp. 1249–1252. IEEE.
10. Liu, S., Cai,W.,Wen, L. & Feng, D. (2011). Volumetric Congruent Local Binary Patterns for 3D Neurological Image Retrieval in The 26th International Conference on Image and Vision Computing New Zealand (IVCNZ) Delmas, P., Wuensche, B. & James, J. (Eds.) (IVCNZ), pp. 272–276. 94 REFERENCES
11. Liu, S. et al. (2010). A Robust Volumetric Feature Extraction Approach for 3D Neuroimaging Retrieval in The 32nd Annual International Conference of the IEEE Engineering in Medicine and Biology Society (EMBC), pp. 5657–5660. IEEE.

12. Liu, S. et al. (2010). Localized Multiscale Texture based Retrieval of Neurological Image in The 23rd IEEE International Symposium on Computer-Based Medical Systems (CBMS), pp. 243–248. IEEE.

13. Liu, S. et al. (2011). Localized Functional Neuroimaging Retrieval using 3D Discrete Curvelet Transform in IEEE International Symposium on Biomedical Imaging: From Nano to Macro (ISBI), pp. 1877–1880. IEEE.

14. Liu, S. et al. (2013). A Supervised Multiview Spectral Embedding Method for Neuroimaging Classification in The 20th IEEE International Conference on Image Processing (ICIP), pp. 601–605. IEEE.

15. Liu, S. et al. (2013). Localized Sparse Code Gradient in Alzheimers Disease Staging in The 35th Annual International Conference of the IEEE Engineering in Medicine and Biology Society (EMBC), pp. 5398–5401. IEEE.

16. Liu, S., et al. (2014). Multi-channel neurodegenerative pattern analysis and its application in alzheimers disease characterization. *Computerized Medical Imaging and Graphics*, *38*, 436–444. issn: 0895-6111.

17. Liu, S. et al. (2016). Cross-View Neuroimage Pattern Analysis for Alzheimers Disease Staging. Frontiers in Aging Neuroscience.

18. Minoshima, S., Frey, K. A., Koeppe, R. A., Foster, N. L., & Kuhl, D. E. (1995). A diagnostic approach in alzheimers disease using three-dimensional stereotactic surface projections of fluorine-18-FDG PET. *Journal of Nuclear Medicine*, *36*, 1238–1248.

19. Ngiam, J., Coates, A., Lahiri, A., et al. (2011). On Optimization Methods for Deep Learning in The 28th International Conference on Machine Learning, pp. 265–272.

20. Poultney, C., Chopra, S. & LeCun, Y. (2006). Efficient Learning of Sparse Representations with An Energy-Based Model in Advances in Neural Information Processing, pp. 1137–1144.

21. Singh, N., Wang, A., Sankaranarayanan, P., Fletcher, P. & Joshi, S. (2012). Genetic, Structural and Functional Imaging Biomarkers for Early Detection of Conversion from MCI to AD in Medical Image Computing and Computer-Assisted Intervention (MICCAI) Ayache, N., Delingette, H., Golland, P. & Mori, K. (Eds.), pp. 132–140, vol. 7510. Springer: Berlin, Heidelberg. ISBN: 978-3-642-33414-6.

22. Suk, H.-I., Lee, S., & Shen, D. (2013). Latent feature representation with stacked auto- encoder for AD/MCI diagnosis. *Brain Structure and Function*, *220*, 841–959.

23. Toews, M., III, W. W., Collins, D. L., & Arbel, T. (2010). Feature-based morphometry: Discovering group-related anatomical patterns. *NeuroImage*, *49*, 2318–2327. ISSN: 1053-8119. 95 CHAPTER 4. ENCODING THE NEURODEGENERATIVE FEATURES.

24. Zhang, F. et al. (2015). Pairwise Latent Semantic Asociation for Similarity Computation in Medical Imaging. IEEE Transactions on Biomedical Engineering.

Chapter 5
Recognizing the Neurodegenerative Patterns

Neurodegenerative disorders always progress in certain patterns. In the case of Alzheimer's Disease (AD), for example, its pathology starts within hippocampus and entorhinal cortex, and spreads throughout most of the temporal lobe and posterior cingulate, finally reaches the parietal, prefrontal and orbitofrontal regions [10, 11, 35]. Subjects with AD may appear with different patterns, and we could use these patterns to enhance our understanding of the disease and facilitate the diagnosis.[1]

Pattern analysis in neurodegenerative disorders depends on two factors, the feature descriptor and the statistical model. Features, e.g., morphological [4, 18–20, 22, 23, 25, 26, 42], functional [3–5, 30, 32] or multimodal features [7, 25, 29, 37, 39], have been discussed in Chap. 2, Sect. 2.2. A set of novel feature descriptors is introduced in Chap. 4. The statistical models, e.g. single-variant or multi-variant analysis models, are summarized in Chap. 2, Sect. 2.3. When multiple models are used in pattern analysis, we can cross-validate the patterns derived from different statistical models, and identify the most essential brain regions associated with the disease. We refer to each statistical model as a '**Channel**'. When multimodal features are available in pattern analysis, we can view the brain from different perspectives, thus obtain a more comprehensive observation in the disease pathology. We may refer to each feature as '**View**'. This chapter introduces both streams in recognizing the patterns of neurodegenerative disorders. Section 5.1 reports the methods and findings by channel-based pattern analysis, and Sect. 5.2 focuses on the view-based pattern analysis.

---

[1]Some content of this chapter has been reproduced with permission from [28, 30].

© Springer Nature Singapore Pte Ltd. 2017
S. Liu, *Multimodal Neuroimaging Computing for the Characterization of Neurodegenerative Disorders*, Springer Theses, DOI 10.1007/978-981-10-3533-3_5

5.1   Channel-Based Pattern Analysis

Channel-based pattern analysis aims to identify the pathology-associated brain regions. It heavily depends on statistical models, such as t-test [13, 24, 28], elastic net [29, 40], and Fisher discriminant ratio [16, 21]. The statistical models can be used in different subject group pairs to identify the differences between different populations, i.e., patients labeled with AD versus normal control (NC), mild cognitive impairment (MCI) versus NC, and AD versus MCI. We refer to each statistical model as a channel, and all the channels are parallel to each other. We argue that multiple channels may have more statistical power than using single channels alone. Therefore, a multi-channel analysis framework was designed to depict the neurodegenerative patterns. In this analysis, we designed 10 parallel channels analyze the neurodegenerative patterns of AD and MCI, and they together helped us to identify the most important brain regions to distinguish different pathologies. The ADNI PET subset was used to test the proposed multi-channel pattern analysis method, with mean Index (M-IDX) [3, 4] of each brain region being used as the feature.

5.1.1   Single-Channel Definition

Table 5.1 shows the combination of the group pairs and the statistical models, including the two-one-sided-test (TOST), support vector machine (SVM) and elastic net (EN).

Two One-Sided Test Channels Three TOST-based channels were designed for AD versus NC (CH 1), MCI versus NC (CH 4), and AD versus MCI (CH 7), respectively. TOST is a classic statistical test, arguably the most commonly used test, in pattern analysis [4, 13]. In TOST, a null hypothesis of equivalence, H_0, will be rejected if the *p-value* of the test statistic is smaller than a user-defined threshold. In this study, we set the threshold as 0.05. The significant feature inputs that passed the test were recorded in a vector as the analysis result.

Table 5.1 The channels defined in the proposed multi-channel analysis framework. Digits in this table represent the channel index. Table reproduced with permission from [28]

Group pair	TOST	SVM	EN
AD versus NC	1	2	3
MCI versus NC	4	5	6
AD versus MCI	7	8	9
AD versus MCI versus NC	n/a	n/a	10

Support Vector Machine Channels Another three SVM-based channels were designed for the same group pairs as in TOST, i.e., AD versus NC (CH 2), MCI versus NC (CH 5), AD versus MCI (CH 8). SVM is the most widely used technique in classification and regression analysis [6]. SVM separate two groups of subjects with a maximum margin, which appears as a hyper-plane in the feature space. The slopes of the hyper-plane on each feature dimension reflect the importance of the feature. Given a set of observations (x_m, y_m), $m = 1, \ldots, M$ where $x_m \in R^N$ is the feature vector, and $y_m \in \{-1, 1\}^M$ is the label, SVM solves the problem:

$$\widehat{w} = \arg \min_w \frac{1}{2} w^T w + C \sum_{m=1}^{M} \epsilon_m \qquad (5.1)$$

$$\text{subject to } y_m(w^T w + b) \geq 1 - \epsilon_m; \epsilon_m \geq 0.$$

where C is a parameter that determines the trade-off between increasing the margin-size and ensuring the training samples lie on the correct side. In this study, SVM was implemented using the LIBSVM [6]. Then 10-fold cross validation was first carried out to obtain the best estimate of C, by exhaustively searching values in [1, 100], and the optimal value was 8. We then solved the problem in Eq. 5.1 using CVX (http://cvxr.com/cvx). Only the regions with weights in the upper quartile of the entire distribution were selected.

Elastic Net Channels The last four channels were based on EN, which is a well established algorithm in feature selection. EN solves the problem:

$$\widehat{\beta} = \arg \min_\beta ||y - X\beta||^2 + \lambda_1 ||\beta|| + \lambda_2 ||\beta||^2 \qquad (5.2)$$

where y is the label vector of M observations, X is the matrix of M feature vectors, $X = \{x_1, \ldots, x_m, \ldots, x_M\}^T$, λ_1 and λ_2 are the regularization parameters, and β is the weight vector with same dimension as x_m. EN introduces both l_1 and squared l_2 penalty of β, which encourage the grouping of feature variables and meanwhile remove the limitation on the number of selected features. This characteristic makes EN an ideal candidate in the high-sample-size low-dimension scenarios, where the datasets have a large number of observations and a small number of features, or correlated features. In this analysis, for example, the number of patients in the ADNI PET subset is much larger than the number of features/brain regions, and the brain regions were highly correlated. Another superior characteristic of EN, compared to TOST and SVM, is that it can select the features in multi-categorical regression. Therefore, we designed four EN-based channels, i.e., AD versus NC (CH 3), MCI versus NC (CH 6), AD versus MCI (CH 9), and AD versus MCI versus NC (CH 10). In this study, the weight of λ_1 and λ_2 were equally set to 0.5.

Figure 5.1 illustrates the patterns from all channels expected for those from the MCI versus NC pairs. Hippocampus, posterior cingulate gyrus and parietal lobe were detected in CH 1–3 as the key regions to distinguish AD from NC. In the experiments on CH 4–6, only few regions were able to differentiate MCI and NC. CH 7–9 had

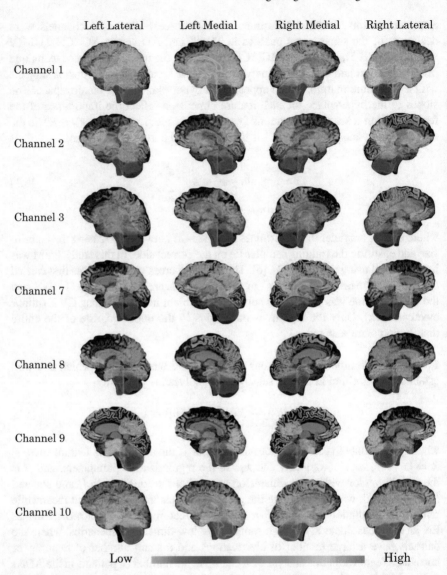

Fig. 5.1 The projected PET patterns derived from individual channels. The *color bar* indicates the statistical power. Figure reproduced with permission from [28]

strong agreement on parietal lobe, which showed a measurable difference between AD and MCI. CH 10 identified 18 regions spreading across the brain, including hippocampus, amygdala, cingulate gyrus, and parietal lobe. These findings were consistent with previous findings in AD and MCI. We also found that temporal pole, brainstem, subgenual and pre-subgenual frontal cortex were strongly associated with AD.

5.1.2 Multi-Channel Voting

The patterns derived from these 10 parallel channels provided complementary information to each other. We adopted a voting approach to integrate them and evaluate the brain regions' statistical power in AD and MCI characterization, as in Eqs. 5.3 and 5.4:

$$\{V := v_n, n = 1, \ldots, N\}$$
$$v_n = \sum_{c=1}^{C} sgn(v_{c,n}) \tag{5.3}$$

$$sgn(v_{c,n}) = \begin{cases} 1, & \text{if } v_{c,n} \neq 0 \\ 0, & \text{if } v_{c,n} = 0 \end{cases} \tag{5.4}$$

where $v_{n,c}$ is the output of the nth region in the cth channel, $N = 83$ is the number o brain regions of interest (ROIs), and $C = 10$ is the number of channels. This multi-channel scheme let individual channels to vote on each brain region, thus assign higher values to the brain regions recognized by more channels [24, 28]. In other words, multi-channel method could depict the statistical power of individual regions on a more objective basis, for more votes on a ROI means stronger agreement among the channels in characterizing the disorders. Table 5.2 lists the voting scores of the 83 brain ROIs defined in the ICBM_152 template [31], and Fig. 5.2 illustrates the integrated pattern.

Multi-channel voting merged the brain ROIs detected by all the individual channels, thereby showing a more spreading pattern. The fused pattern, as illustrated in Fig. 5.2, involved more brain ROIs when compared to the patterns of individual channels. The votes of each brain ROI reflected its statistical power, therefore the pattern's contrast was also remarkably increased.

5.1.3 Multi-Channel Analysis in Neuroimaging Retrieval

As shown in Table 5.2, the proposed multi-channel method identified the disorder sensitive regions, but excessively resulted 67 brain ROIs with at least 1 vote from these single channels. For example, 19 out of the 67 brain regions were detected by single channels only. These regions were suspiciously the 'lucky results' caused by biased group size or the drawbacks of the pattern analysis algorithm, and could potentially damage characterization performance. To optimize the multi-channel analysis in the characterization of AD, we proposed an iterative feature selection approach, which tested a set of feature selection criteria and identified the optimal subset of the brain ROIs in the integrated pattern. All of the brain regions were sorted in ascent order based on their votes. We then carried out a series of retrieval experiments to characterize AD and MCI using the subsets of the brain ROIs, which were selected

Table 5.2 Scores of the 83 brain ROIs by multi-channel voting. Left column: the region index in the ICMB_152 template. Middle column: the brain structure labels. Right column: the multi-channel voting scores. Table reproduced with permission from [28]

Index	Region label (Right, Left)	Score
(1, 2)	Hippocampus (R, L)	(6, 5)
(3, 4)	Amygdala (R, L)	(3, 6)
(5, 6)	Anterior temporal lobe (R, L)	(1, 0)
(7, 8)	Anterior temporal lobe, lateral part (R, L)	(0, 3)
(9, 10)	Parahippocampal and ambient gyri (R, L)	(4, 1)
(11, 12)	Superior temporal gyrus, posterior part (R, L)	(2, 4)
(13, 14)	Middle and inferior temporal gyrus (R, L)	(6, 4)
(15, 16)	Fusiform gyrus (R, L)	(2, 2)
(17, 18)	Cerebellum (R, L)	(1, 1)
19	Brainstem (unpaired)	4
(20, 21)	Insula (R, L)	(0, 0)
(22, 23)	Lateral remainder of occipital lobe (R, L)	(0, 3)
(24, 25)	Cingulate gyrus, anterior part (R, L)	(0, 2)
(26, 27)	Cingulate gyrus, posterior part (R, L)	(4, 5)
(28, 29)	Middle frontal gyrus (R, L)	(2, 0)
(30, 31)	Posterior temporal lobe (R, L)	(2, 1)
(32, 33)	Inferiolateral remainder of parietal lobe (R, L)	(6, 5)
(34, 35)	Caudate nucleus (R, L)	(1, 1)
(36, 37)	Nucleus accumbens (R, L)	(2, 2)
(38, 39)	Putamen (R, L)	(2, 4)
(40, 41)	Thalamus (R, L)	(1, 0)
(42, 43)	Pallidum (R, L)	(1, 2)
44	Corpus callosum (unpaired)	3
(45, 46)	Lateral ventricle - (apart temporal horn) (R, L)	(0, 4)
(47, 48)	Lateral ventricle, temporal horn (R, L)	(3, 3)
49	Third ventricle (unpaired)	0
(50, 51)	Precentral gyrus (R, L)	(1, 1)
(52, 53)	Straight gyrus (R, L)	(0, 0)
(54, 55)	Anterior orbital gyrus (R, L)	(2, 1)
(56, 57)	Inferior frontal gyrus (R, L)	(1, 1)
(58, 59)	Superior frontal gyrus (R, L)	(0, 0)
(60, 61)	Postcentral gyrus (R, L)	(2, 1)
(62, 63)	Superior parietal gyrus (R, L)	(4, 4)
(64, 65)	Lingual gyrus (R, L)	(3, 2)
(66, 67)	Cuneus (R, L)	(0, 1)
(68, 69)	Medial orbital gyrus (R, L)	(0, 2)
(70, 71)	Lateral orbital gyrus (R, L)	(1, 2)

(continued)

Table 5.2 (continued)

Index	Region label (Right, Left)	Score
(72, 73)	Posterior orbital gyrus (R, L)	(2, 2)
(74, 75)	Substantia nigra (R, L)	(2, 5)
(76, 77)	Subgenual frontal cortex (R, L)	(6, 6)
(78, 79)	Subcallosal area (R, L)	(3, 4)
(80, 81)	Pre-subgenual frontal cortex (R, L)	(6, 1)
(82, 83)	Superior temporal gyrus, anterior part (R, L)	(1, 2)

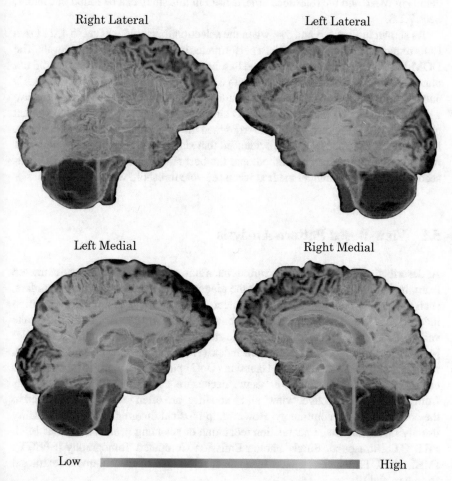

Fig. 5.2 The projected PET patterns derived from the integrated channel. The *color bar* indicates the statistical power. Figure reproduced with permission from [28]

in a 'many to few' manner based on the regions' votes. For example, regions with votes larger than 1 were selected in first experiment; regions with votes larger than 2 selected in the second experiment; and so on.

The selected ROI features were compared to the whole brain features (Baseline, with all ROI features), disorder-oriented mask features (DOM, with 40 features selected based on prior knowledge [28]), TOST-selected (TOST), SVM-selected (SVM), and EN-selected (EN) ROI features in the characterization of AD and MCI. The performance was evaluated using the mean average precision (MAP) metric with soft relevance criteria to evaluate the relevance of AD, MCI and NC subjects. More details of MAP and the relevance criteria used in this study can be found in Chap. 7, Sect. 7.1.3.

As shown in Figs. 5.3 and 5.4, when the selection threshold was set to 4, 21 brain ROIs remained and achieved the best performance in both AD and MCI retrievals. The DOM-based method outperformed the baseline method on AD by 0.3%, but did not match baseline performance on MCI with a decrease of 1.9%. Such conflicts on AD and MCI might be raised by too many brain ROIs being used in DOM. It might have a negative impact on the retrieval of AD subjects due to the excessive involvement of MCI-sensitive regions. We further examined the single channel analysis methods, i.e., TOST, SVM and EN. It was evidenced that single channel methods were not as robust as the multi-channel method, and the best single channel method, EN, was second to the multi-channel method when the vote threshold was set to 4.

5.2 View-Based Pattern Analysis

As described in Chap. 2, Sect. 2.1, various quantitative measurements can be extracted from the PET and MRI data to assist the diagnosis of neurodegenerative disorders, such as cortical thickness [12], grey matter volume (GMV) [1], local gyrification index (LGI) [36], solidity (SLD) and convexity (CNV) [25, 26], standard uptake value (SUV) [9, 17], cerebral metabolic rate of glucose consumption (CMRGlc) [4, 38], hypo-metabolic convergence index (HCI) [8], mean index (M-IDX) [3], z-scores [32], and difference-of-Gaussian (DoG) parametric maps [5]. We refer to each type of measurement as a 'view', because it allows us to view the patients from a specific perspective. 'view' and 'modality' are often interchangeably used in the computer vision community. However, in medical imaging domain, a modality usually means an image acquisition technique or scanning protocol, such as MRI, PET, CT, Ultrasound, Single Photon Emission Computed Tomography (SPECT), fMRI, and DTI, whereas a view means a specific type of measurements extracted from a modality.

To unearth the buried treasure in multi-view neuroimaging data, researchers have carried out many studies on multi-view feature fusion and related applications in neurological disorders [2, 7, 13–15, 25, 27, 34, 41, 43]. The aforementioned studies demonstrated that multi-view features provide complementary information to individual single-view features, thus lead to a better performance in the recognizing

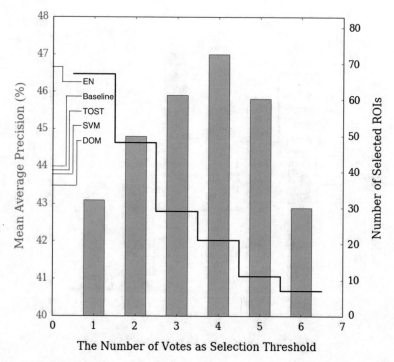

Fig. 5.3 The MAP (%) for AD retrieval compared to DOM, Baseline, TOST, SVM and EN approaches. Figure reproduced with permission from [28]

the neurological patterns. In other words, multi-view features create a synergy in the characterization of AD and MCI, However, it is not known what causes such synergy, and how to quantitatively evaluate the synergy between different views. As discussed in previous section, it is evident that there is marked differences between patterns derived from different views. We hypothesized that the variability of neurodegenerative patterns might have a strong correlation with the synergy. To test our hypothesis, we developed a cross-view pattern analysis framework to measure the synergy of the multi-view features. Figure 5.5 illustrates the work-flow of the proposed cross-view framework. The ADNI MRI-PET subset was used to evaluate this framework. Details about the dataset and pre-processing procedures can be found in Chap. 3, Sect. 3.3. Totally 9 views of features were tested in this study, including 4 MRI features and 5 PET features. The features were all ROI-based, thus have equal dimensions. Chapter 2, Sect. 2.2 gives the details of these features. Experiments on single-view and cross-view pattern analysis methods were carried out on these views, and the combinations of these views were evaluated in the characterization of AD, MCI and normal control (NC) subjects using MK-SVM.

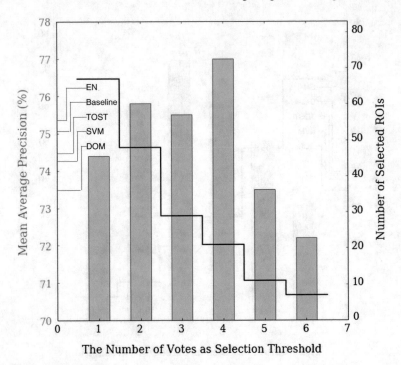

Fig. 5.4 The MAP (%) for MCI retrieval compared to DOM, Baseline, TOST, SVM and EN approaches. Figure reproduced with permission from [28]

5.2.1 Single-View Pattern Analysis

The features extracted from the T1-weighted MRI data include Grey Matter Volume (GMV), Local Gyrification Index (LGI), Convexity Ratio (CNV) and Solidity Ratio (SLD). The features extracted from FDG-PET data include Mean Index (M-IDX), Fuzzy Index (F-IDX) and three Difference-Of-Gaussian features (DoG-M, DoG-C, DoG-Z). For each of the nine views, we carried out analysis of variance (ANOVA) on the three population groups, i.e., AD, MCI and NC, to test the null hypothesis that all groups were random samples of the same population. Given a view, P, its p-values of ANOVA, $P = \{P(1), P(2), \ldots, P(K)\}$, reflected the discriminating power of individual ROIs in this view. We further transformed these p-values to non-negative valued weights which were positively correlated to the ROI p-values, as in Eq. (5.5):

$$P'(i) = \exp\left(-\frac{P(i)^2}{2\sigma^2}\right) \tag{5.5}$$

where σ is the bandwidth parameter which controls how fast $P'(i)$ falls off with the $P(i)$. If $P(i)$ is significantly small, then $P'(i)$ is close to the maximum value 1; and if $P'(i)$ is greater than σ, then $P'(i)$ will plummet to 0. Following the convention, the

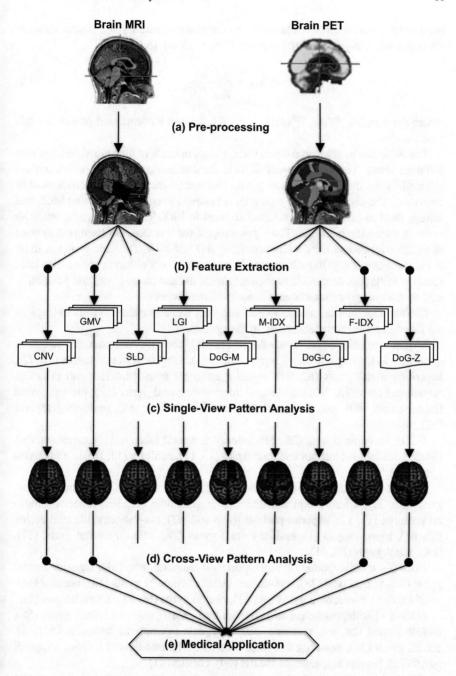

Fig. 5.5 The cross-view pattern analysis framework, which deduces the pathological patterns from different views and explores the correlations between them. Figure reproduced with permission from [30]

bandwidth σ was set to 0.05. In order to compare the patterns of different views in the following analysis, we further normalized P' as Eq. (5.6):

$$P''(i) = \frac{P'(i)}{\sum_K P'(j)} \tag{5.6}$$

where the weights, $P''(1), P''(2), \ldots, P''(K)$, formed a normalized pattern for this specific view.

The ROIs can be categorized into three groups in terms of their consistency across different views. The first group of ROIs is the disease-spared ROIs, which are not affected by the disorders, appearing with low discriminating power across most of the views. The cerebellum, for example, is believed to spared by AD and MCI, and always used to calibrate the metabolism rates in PET. The second group of ROIs is the disease-affected ROIs. The hippocampus, for instance, has been widely used as an effective biomarker for characterizing AD and MCI. The third group of ROIs is the view-specific ROIs, which vary in *p-values* across different views. The third group of ROIs encode the different effects of the disease on the brain, and potentially lead to the synergy or interference among different views.

The brain regions deduced from PET and MRI are summarized as follows (regions are listed in ascent order according to their *p-values*):

M-IDX - inferiolaterial remainder of parietal lobe (32, 33), posterior cingulate gyrus (26, 27), amygdala (3, 4), hippocampus (1), subgenual frontal cortex (76, 77), superior parietal gyrus (62, 63), superior temporal gyrus posterior part (11, 12), subcallosal area (78, 79), middle and inferior temporal gyrus (13), pre-subgenual frontal cortex (80), middle frontal (28), caudate nucleus (34), posterior temporal (30).

F-IDX - middle frontal (28, 29), anterior temporal lobe (6, 8), caudate nucleus (34, 35), middle and interior temporal gyrus (13, 14), fusiform (15, 16), lateral orbital gyrus (70).

DoG-M - amygdala (3, 4), subgenual frontal cortex (76, 77), posterior cingulate gyrus (26, 27), inferiolateral remainder of parietal lobe (32, 33), superior temporal posterial (11, 12), superior parietal gyrus (62, 63), pre-subgenual frontal cortex (80, 81), hippocampus (1), middle frontal gyrus (28), inferior frontal gyrus (57), postcentral gyrus (60, 61).

DoG-C - middle frontal (28, 29), posterior temporal lobe (31), superior frontal gyrus (58), subcallosal (78), middle and inferior temporal gyrus (13), anterior temporal (5, 6, 8), postcentral gyrus (60, 61), orbital gyrus (69, 71), hippocampus (1).

DoG-Z - parahippocampal and ambient gyri (9, 10), superior frontal gyrus (58), middle frontal (28, 29), putamen (39), amygdala (4), anterior temporal (5, 6, 8), orbital gyrus (70), posterior temporal lobe (30, 31), middle and inferior temporal gyrus (13), insula (21), superior frontal (60), cuneus (67).

GMV - hippocampus (1, 2), amygdala (3, 4), caudate nucleus (34, 35), middle and inferior temporal gyrus (13, 14), parahippocampal and ambient gyrus (9, 10), ventricles (45, 46, 47, 48, 49), anterior temporal lobe (5, 6), anterior cingulate gyrus

(24, 25), lateral remainder of occipital (22), fusiform (15, 16), superior frontal gyrus (58).

LGI - posterior temporal lobe (30), caudate nucleus (34, 35), ventricles (45), superior frontal gyrus (58, 59), putamen (39), cuneus (66, 67), anterior cingulate gyrus (25), lingual gyrus (65).

CNV - ventricles (45, 46, 47, 48), posterior temporal lobe (30), caudate nucleus (35), lingual (65), superior frontal gyrus (58), lateral remainder of occipital (22), thalamus (41), cerebellum (18), cuneus (66).

SLD - amygdala (3, 4), lingual (64, 65), middle and interior temporal gyrus (13, 14), postcentral gyrus (60, 61), cuneus (66, 67), parahippocampal (9), lateral remainder of occipital (22), superior parietal gyrus (63).

Figure 5.6 shows the back-projection of single-view patterns derived from PET and MRI onto the ICBM_152 brain template [31]. The color bar indicates the *p-values* of the ROIs in each view. Note that the ventricles and corpus callosum are not displayed here. Based on patterns derived from MRI, we found that many ROIs of the brain were spared by the disease, including the insula, brain stem, corpus callosum, and parts of the frontal lobe, parietal lobe and subcortical regions. The disease-affected regions were consistent to the established knowledge, including ventricles, middle and inferior temporal lobe and limbic gyrus. We also noticed there was a strong agreement across most views on the occipital lobe (lateral part, lingual and cuneus) and frontal lobe (superior part), which were less reported in previous studies. There were also a number of view-specific ROIs, i.e., GMV detected the hippocampus, parahippocampal and ambient gyrus, and amygdala; CNV detected two particular ROIs, the cerebellum and the thalamus, although these two structures were usually considered spared by AD; SLD detected two unique ROIs in the parietal lobe (superior and post-central parts). In addition to the well known temporal lobe and limbic gyrus detected by MRI, the PET patterns also detected more frontal regions (subgenual, orbital, inferior, middle and superior parts) and parietal regions (post-central and superior parts). These regions are believed to be affected later in the course of AD and MCI, posterior to the hippocampus, entorhinal cortex, temporal regions and posterior cingulate [11]. This finding suggested that the frontal and parietal lobes were essential biomarkers in staging the AD and MCI patients, and we potentially can detect the functional changes prior to the structural changes in these regions. Compared to MRI views, PET views were less sensitive to pathological changes in the occipital lobe, where only the cuneus was detected by the DoG-M and DoG-Z. The patterns of M-IDX and DoG-M were larger than the other views, both covering the inferiolateral parietal area.

To summarize, we found that there were disease-spared regions, such as insula, brain stem, corpus callosum, verified by both PET and MRI. MRI views were capable of capturing the brain structural changes on temporal lobe, limbic gyrus, ventricles, and part of the occipital lobe, which were usually affected later in the course of the disease. The PET views, on the other hand, were able to detect both the early and late functional anomalies, thereby involving more ROIs in their patterns than the MRI views, especially in the frontal and parietal lobes. Furthermore, some ROIs were only perceivable in certain views, and led to the distinct patterns. The varying patterns of

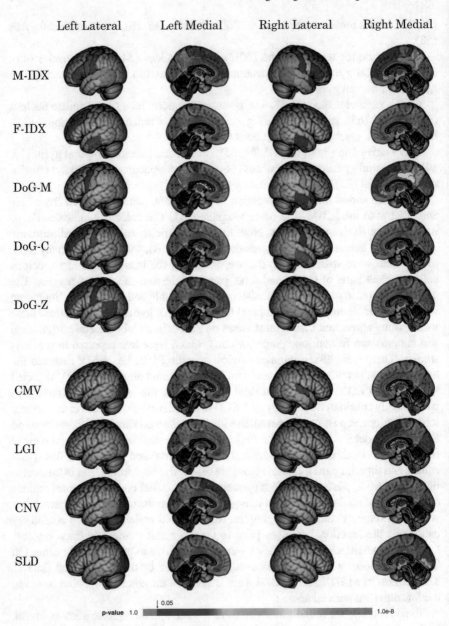

Fig. 5.6 Projection of the weighted ROIs derived from each view onto the ICBM_152 template. Figure reproduced with permission from [30]

the views indicated that the pathology may appear differently on the brain and no single-view was able to capture all the pathological changes.

5.2.2 Cross-View Pattern Analysis

Figure 5.5 - Step (d) illustrates the cross-view pattern analysis method. The goal of this step was to compare the patterns and quantitatively analyze the complementary information between them. We first paired up these $M = 9$ views, which lead to $M \times (M - 1)/2 = 36$ combinations of views, then quantitatively analyzed each combination based on their patterns. Assuming P'' and Q'' are the patterns of two views their affinity, $A(P, Q)$, is defined as in Eq. 5.7:

$$A(P, Q) = \exp\left(-\frac{1}{2\pi}(\overbrace{\sum_K P''(i) \log \frac{P''(i)}{Q''(i)}}^{D_{KL}(P||Q)} + \overbrace{\sum_K Q''(i) \log \frac{Q''(i)}{P''(i)}}^{D_{KL}(Q||P)}) \right) \quad (5.7)$$

where $D_{KL}(P||Q)$ is the Kullback–Leibler (KL) divergence from P to Q, and $D_{KL}(Q||P)$ is the KL divergence from Q to P. Notice that KL divergence is a non-symmetric measure of difference between P and Q, and does not satisfy the symmetry condition to be used as a similarity metric. Therefore, we actually measured the affinity between two views based on their mutual divergence. $A(P, Q) = 0$ if $P'' = Q''$.

The affinity measurements of all pairs formed an affinity matrix A. We computed the clustering of them to see how they were related, based on the symmetric normalized Laplacian matrix (L) of A [33], as in Eq. (5.8):

$$L = I - D^{-1/2}AD^{-1/2} \quad (5.8)$$

where I is a unit matrix, D is a diagonal matrix whose (i, i)-element was the sum of A's i^{th} row. The patterns can be considered as samples in a K-dimensional space, and the top-k Eigen vectors of L could be stacked in columns to form a new k-dimensional space for the patterns, which allowed us to observe the embedding of the views in a low dimensional space. In this study, k was set to 2 to display the views in a 2-dimensional space.

Table 5.3 shows the KL divergence ($D_{KL}(Col||Row)$) from the column item (Col) for these 9 views to the row item (Row). Both PET and MRI views had a low mean KL divergence, i.e., 16.6 of PET views compared 17.8 of MRI views. However, the mean KL divergence from MRI views to PET views ($D_{KL}(MRI||PET) = 47.37$) was markedly higher than that from PET views to MRI views ($D_{KL}(PET||MRI) = 20.87$). These findings indicated that the views of the same modality usually look more similar than those of different modalities. GMV, for example, showed the dis-

Table 5.3 The cross-view KL divergence ($D_{KL}(Col||Row)$) from the column item (*Col*) to the row item (*Row*). Table reproduced with permission from [30]

$D_{KL}(Col\|\|Row)$		GMV	LGI	CNV	SLD	M-IDX	F-IDX	DoG-M	DoG-C	DoG-Z
MRI	GMV	0	32.4	33.7	25.0	5.2	23.2	12.6	28.4	16.8
	LGI	6.8	0	16.3	28.3	3.4	21.7	9.2	28.2	22.3
	CNV	2.3	20.4	0	38.7	4.0	19.4	28.4	45.9	46.2
	SLD	4.0	32.9	25.4	0	5.4	39.2	11.7	31.8	14.3
PET	M-IDX	48.1	48.9	51.8	49.8	0	39.1	6.0	45.3	34.6
	F-IDX	31.1	43.0	50.3	55.7	13.0	0	18.9	18.0	9.9
	DoG-M	50.0	50.2	56.9	47.6	0.4	42.7	0	55.1	35.1
	DoG-C	39.3	52.3	40.6	61.5	6.9	27.9	18.6	0	13.3
	DoG-Z	33.5	43.8	45.1	47.9	9.2	27.5	19.3	4.1	0

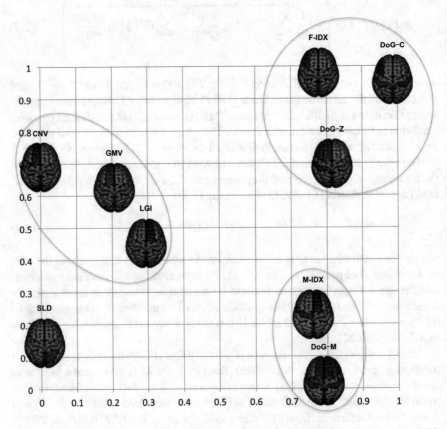

Fig. 5.7 Clustering results of 9 views in 2D space. Figure reproduced with permission from [30]

parity in inter-modal and intra-modal KL divergences. It had a small divergence from other MRI views $(D_{KL}(LGI||GMV) = 6.83;\ D_{KL}(CNV||GMV) = 2.31;$ $D_{KL}(SLD||GMV) = 3.99)$, but large divergence from PET views $(D_{KL}(M\text{-}IDX$ $||GMV)=48.09;\ D_{KL}(F\text{-}IDX||GMV)=31.13;\ D_{KL}(DoG\text{-}M||GMV) = 49.95;$ $D_{KL}(DoG\text{-}C||GMV) = 39.28;\ D_{KL}(DoG\text{-}Z||GMV) = 33.53)$. It was also evident that the MRI views always gain more information from the PET views than the other way around, e.g., $D_{KL}(M\text{-}IDX||CNV) = 51.80$ was 12 times larger than $D_{KL}(CNV||M\text{-}IDX) = 4.01; D_{KL}(DoG\text{-}M||LGI) = 50.15$ was 5 times larger compared to $D_{KL}(LGI||DoG\text{-}M) = 9.21$. except for CNV and DoG-Z both having high divergence from each other. The divergence had a very wide range of values among individual views, from the minimum $D_{KL}(M\text{-}IDX||DoG\text{-}M) = 0.35$ to the maximum $D_{KL}(SLD||DoG\text{-}C) = 61.47$.

Figure 5.7 displays the clustering results of the views in a 2D space based on the distance between them, which is proportional to their mutual divergence. It was evident that MRI and PET views were clearly separated with a large margin. More importantly, the intra-modality views also formed sub-clusters, i.e., two sub-clusters by the MRI views and two sub-clusters by the PET views. The first sub-cluster (C1) in MRI included CNV, GMV, and LGI. They encoded the loss of cortical neurons and the changes of cortical foldings, and all showed strong correlations with the brain cortical atrophy. The second cluster (C2) in MRI had one isolated view only, SLD, which was different from other MRI views with a focus on the fullness of the brain ROIs. The third cluster (C3) contained three PET views, F-IDX, DoG-C and DoG-Z, which could effectively evaluate the consistency of the metabolism levels within ROIs when they were partially hypo-metabolic. The M-IDX and DoG-M, both sensitive to the metabolic activity changes of the brain in the early detection of AD and MCI, belonged to the fourth cluster (C4).

5.2.3   Performance Evaluation

The classification results of individual views were discussed in Chap. 4, Sects. 4.1 and 4.2. Figures 5.8 and 5.9 show the classification results of 36 combinations of features in the same classification task, including 6 intra-MRI combinations, 10 intra-PET combinations and 20 inter MRI-PET combinations. The best performance results are highlighted in bold-face.

Feature combinations, with only few exceptions, outperformed the single-view features in the characterization of AD and MCI with marked improvements. Same as the single-view features, there was no combination superior to all the other combinations in every aspect. The combination of CNV and DoG-M achieved the best precision for NC (65.94%) and the best overall sensitivity (93.69%), which were 12.64 and 11.42% higher than the best single-view performance. The combination of GMV and F-IDX had the best performance on MCI classification (70.89%) and AD classification (80.56%), which were 6.83 and 12.92% better than the best single-view performance. The combination of F-IDX and DoG-M, dramatically

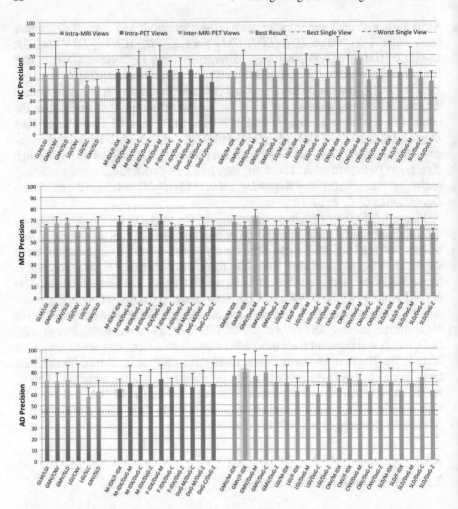

Fig. 5.8 The classification performance on NC, MCI, and AD patients, using the combinations of the features. *Blue* the MRI–MRI combination; *Red* the PET–PET combination; *Purple* the MRI–PET combinations

improved the overall accuracy from 56.49% using F-IDX alone to 67.37%. The overall specificity peaked at 69.92%, which was achieved by CNV and DoG-C with an increase of 6.25% compared to the best single-view specificity.

It was evident inter-MRI-PET combinations most likely gave better results than the intra-PET and intra-MRI combinations. The only exception was the overall accuracy, which was achieved by the intra-PET combination of F-IDX and DoG-M. However, if we separated the views into different sub-clusters as in Fig. 5.7, it was valid that the best combinations always contained two views from different clusters, for example,

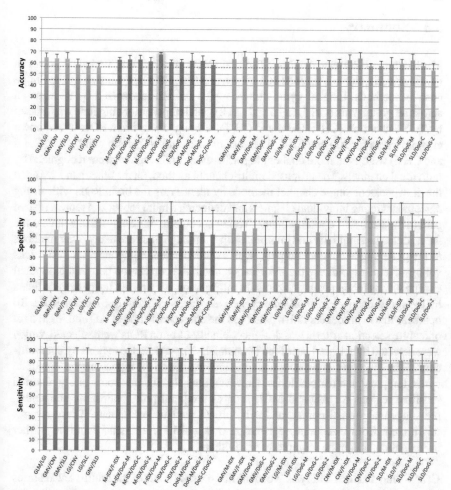

Fig. 5.9 The overall classification performance using the combinations of the features. *Blue* the MRI–MRI combination; *Red* the PET–PET combination; *Purple* the MRI–PET combinations

CNV (from C1) and DoG-M (from C4) achieved highest NC precision and overall sensitivity; GMV (from C1) and DoG-M (from C4) had best MCI precision; GMV (from C1) and F-IDX (from C3) had highest AD precision; F-IDX (from C3) and DoG-M (from C4) attained highest overall accuracy; and CNV (from C1) and DoG-C (from C3) achieved best specificity.

5.3   Summary

Multi-channel Pattern Analysis The brain ROIs have varying statistical power in the characterization of AD and MCI, which are difficult to be evaluated by the Single channel methods. Lingual gyrus, for example, was not recognized by SVM as an across all group pairs (CH 2, CH 5 and CH 8), and left posterior cingulate was missed by TOST channels (CH 1, CH 4 and CH 7), although these brain ROIs were known as biomarkers for AD and MCI. The case of MCI and NC group pair (CH 4, CH 5 and CH 6) were even more challenging, where few brain regions were recognized by individual channels. The proposed multi-channel method, on the other hand, avoided the drawbacks of single channel methods by identifying the brain ROIs with strong agreement among individual channels and deducing a more robust and representative pattern. We adopted a multi-channel voting scheme in this analysis to evaluate the ROIs, dut to its simplicity and interpretability, although there are many other methods to integrate single channel outputs, such as independent component analysis. This voting scheme did not use the information about the correlation between individual channels, thus might result in redundant information. As a result, the pattern deduced by the multi-channel method was excessively large, containing 67 brain ROIs. To optimize this neurodegenerative pattern in the characterization of AD and MCI, a set of feature selection criteria was designed. Finally 21 brain ROIs were selected, which were detected by at least 4 out of the 10 channels. These detected regions by our multi-channel method were consistent with previous studies on the hippocampus, cingulate gyrus, temporal and parietal lobes. However, they also include some rarely reported sensitive regions for AD and MCI, such as temporal pole, brainstem, subgenual, pre-subgenual frontal cortex, and sub-cortical nucleus accumbens. These regions might stem from inaccurate registration or acquisition artifacts, therefore further investigations might be required to verify them.

Cross-View Pattern Analysis We identified 4 sub-clusters of the multi-view features based on their mutual divergence, which is an effective measure to quantize the difference in the features. It was evident that the best classification results were always achieved by the combination of inter-cluster views. However, it was still not clear whether mutual divergence is a statistically meaningful indicator to predict the performance of any combinations of features. In other words, what could we expect from the multi-view features when they are combined in the characterization of AD and MCI?

To answer the above mentioned question, we looked into the correlation between the joint performance of the features combinations and the performance metrics of individual features. E_{joint}, E_{high} and E_{low} represent the joint performance, the higher performance and lower performance of the features. Tables 5.4 and 5.5 show the Pearson's Correlation Coefficients (ρ) and the p-$values$, which are significant if less than 0.05. It was evident that the multi-view classification performance was strongly correlated to the performance of individual views, and largely determined by the view with the higher performance.

Table 5.4 Pearson's correlation coefficient (ρ) between the precision of NC, MCI and AD and the mutual divergence of multi-view features. Table reproduced with permission from [30]

E_{joint}	NC		MCI		AD	
	ρ	p-value	ρ	p-value	ρ	p-value
E_{high}	0.43	3.1E−17	0.43	2.5E−17	0.62	6.6E−39
E_{low}	0.31	2.1E−09	0.35	1.6E−11	0.40	4.1E−15
D_{mutual}	0.00	1.0E+00	−0.03	5.6E−01	−0.07	1.6E−01
D_{high}	0.11	4.0E−02	0.02	6.5E−01	−0.03	5.4E−01
D_{low}	−0.12	2.7E−02	−0.08	1.3E−01	−0.09	8.1E−02

Table 5.5 Pearson's correlation coefficient (ρ) between the multi-class classification performance and the mutual divergence of multi-view features. Table reproduced with permission from [30]

E_{joint}	Accuracy		Specificity		Sensitivity	
	ρ	p-value	ρ	p-value	ρ	p-value
E_{high}	0.66	8.2E−47	0.51	2.3E−25	0.26	3.4E−07
E_{low}	0.45	4.7E−19	0.21	5.0E−05	0.13	1.5E−2
D_{mutual}	−0.10	4.8E−02	0.09	9.8E−02	−0.07	2.2E−01
D_{high}	−0.01	9.0E−01	0.05	3.6E−01	−0.00	9.5E−01
D_{low}	−0.17	9.8E−04	0.10	6.1E−02	−0.11	4.1E−02

We further examined the correlation between the joint performance E_{joint} and the mutual divergence D_{mutual}, as well as the higher KL divergence D_{high} and lower KL divergence D_{low} between the two views in a combination. If D_{high} and D_{low} are both large, it means that the two views have dramatically different patterns, such as DoG-Z and CNV. If D_{high} is large, but D_{low} is small, it means the pattern of one view covers that of the other, such as M-IDX and CNV. If D_{high} and D_{low} are both small, then the two views are highly similar with large overlaps, such as DoG-M and M-IDX. As shown in Tables 5.4 and 5.5, we didn't observe significant correlations between the mutual divergence D_{mutual} and the joint performance E_{joint}, except for a weak anticorrelation between D_{mutual} and accuracy ($\rho = -0.1, p\_value = 0.048$). The higher KL divergence D_{high} was weakly correlated with the NC precision ($\rho = 0.11, p\_value = 0.040$), but the lower KL divergence D_{low} was anticorrelated with the NC precision ($\rho = -0.12, p\_value = 0.027$), accuracy ($\rho = -0.17$, $p\_value \sim= 0.001$) and sensitivity ($\rho = 0.11, p\_value = 0.041$). Therefore, our answer to the above mentioned question is that, (1) an increase in D_{high} or a decrease in D_{low} likely lead to a better classification performance, (2) large mutual divergence does not necessarily lead to a better performance, since the views might not only create synergy, but also interfere with each other, (3) multi-view performance is largely determined by the performance of individual views, and (4) if one view's pattern covers the other, they tend to perform better than those combinations with highly different or highly similar patterns.

References

1. Ashburner, J., & Friston, J. K. (2000). Voxel-based morphometry - The methods. *NeuroImage*, *11*, 805–821.
2. Atrey, P., Hossain, M. A., El Saddik, A., & Kankanhalli, M. (2010). Multimodal fusion for multimedia analysis: A survey. *Multimedia Systems*, *16*, 345–379. ISSN: 0942-4962.
3. Batty, S., Clark, J., Fryer, T., & Gao, X. (2008). Prototype system for semantic retrieval of neurological PET images, English. In X. Gao, H. Müller, M. Loomes, R. Comley, & S. Luo (Eds.), *Medical imaging and informatics* (Vol. 4987, pp. 179–188). Berlin: Springer. ISBN: 978-3-540-79489-9.
4. Cai, W., et al. (2010). 3D neurological image retrieval with localized pathology-centric CMR-Glc patterns. In *The 17th IEEE International Conference on Image Processing (ICIP)* (pp. 3201–3204). IEEE.
5. Cai, W., et al. (2014). A 3D difference of gaussian based lesion detector for brain PET. In *IEEE International Symposium on Biomedical Imaging: From Nano to Macro (ISBI)* (pp. 677–680). IEEE.
6. Chang, C. C., & Lin, C. J. (2011). LIBSVM: a library for support vector machines. *ACM Transactions on Intelligent Systems Technology (ACM TIST)*, *2*, 27:1–27:27. ISSN: 2157-6904.
7. Che, H., et al. (2014). Co-neighbor multi-view spectral embedding for medical contentbased retrieval. In *IEEE International Symposium on Biomedical Imaging: From Nano to Macro (ISBI)* (pp. 911–914). IEEE.
8. Chen, K., Ayutyanont, N., Langbaum, J. B., Fleisher, A. S., Reschke, C., et al. (2011). Characterizing Alzheimer's disease using a hypometabolic convergence index. *NeuroImage*, *56*, 52–60. ISSN: 1053-8119.
9. Clark, C. M., et al. (2012). Cerebral PET with florbetapir compared with neuropathology at autopsy for detection of neuritic amyloid-œ≤ plaques: A prospective cohort study. *The Lancet Neurology*, *11*, 669–678. ISSN: 1474-4422.
10. Desikan, R., Cabral, H., Hess, C., Dilon, W., et al. (2009). Automated MRI measures identify individuals with mild cognitive impairment and Alzheimer's disease. *BRAIN*, *132*, 2048–2057.
11. Fan, Y., Batmanghelich, N., Clark, C., & Davatzikos, C. (2008). Spatial patterns of brain atrophy in MCI patients, identified vis high-dimensional pattern classificaiton. Predic subsequent cognitie decline. *NeuroImage*, *39*, 1731–1743.
12. Fischl, B., & Dale, A. M. (2000). Measuring the thickness of the human cerebral cortex from magnetic resonance images. *Proceedings of the National Academy of Sciences*, *97*, 11050–11055.
13. Heckemann, R. A., et al. (2011). Automatic morphometry in Alzheimer's disease and mild cognitive impairment. *NeuroImage*, *56*, 2024–2037. ISSN: 1053-8119.
14. Hinrichs, C., Singh, V., Xu, G., & Johnson, S. (2009). MKL for robust multi-modality AD classification. In G. Yang (Ed.), *Medical image computing and computer-assisted intervention (MICCAI)* (Vol. 5762, pp. 786–794). Berlin: Springer.
15. Hinrichs, C., Singh, V., Xu, G., & Johnson, S. (2011). Predictive markers for AD in a multi-modality framework: An analysis of MCI progression in the ADNI population. *NeuroImage*, *55*, 574–589.
16. Khedher, L., Ramirez, J., Gorriz, J., Brahim, A., & Illan, I. (2015). Independent component analysis-based classification of Alzheimers disease from segmented MRI data. In *International Work-Conference on the Interplay between Natural and Artificial Computation* (pp. 78–87). Berlin: Springer.
17. Landau, S. M., et al. (2013). Comparing positron emission tomography imaging and cerebrospinal fluid measurements of beta-amyloid. *Annals of Neurolog*, *74*, 826–836. ISSN: 1531-8249.
18. Liu, S., et al. (2010). A robust volumetric feature extraction approach for 3D neuroimaging retrieval. In *The 32nd Annual International Conference of the IEEE Engineering in Medicine and Biology Society (EMBC)* (pp. 5657–5660). IEEE.

19. Liu, S., et al. (2010). Localized multiscale texture based retrieval of neurological image. In *The 23rd IEEE International Symposium on Computer-Based Medical Systems (CBMS)* (pp. 243–248). IEEE.
20. Liu, S., Cai, W., Wen, L., & Feng, D. (2011). Volumetric congruent local binary patterns for 3D neurological image retrieval. In P. Delmas, B. Wuensche, & J. James (Eds.), *he 26th International Conference on Image and Vision Computing New Zealand (IVCNZ)* (pp. 272–276). IVCNZ.
21. Liu, S., et al. (2011). Generalized regional disorder-sensitive-weighting scheme for 3D neuroimaging retrieval. In *The 33rd Annual International Conference of the IEEE Engineering in Medicine and Biology Society (EMBC)* (pp. 7009–7012). IEEE.
22. Liu, S., et al. (2011). Localized functional neuroimaging retrieval using 3D discrete curvelet transform. In *IEEE International Symposium on Biomedical Imaging: From Nano to Macro (ISBI)* (pp. 1877–1880). IEEE.
23. Liu, S., Cai, W., Wen, L., & Feng, D. (2012). Multiscale and multiorientation feature extraction with degenerative patterns for 3D neuroimaging retrieval. In *The 19th IEEE International Conference on Image Processing (ICIP)* (pp. 1249–1252). IEEE.
24. Liu, S., Cai, W., Wen, L., & Feng, D. (2013). Multi-channel brain atrophy pattern analysis in neuroimaging retrieval. In *IEEE International Symposium on Biomedical Imaging: From Nano to Macro (ISBI)* (pp. 206–209). IEEE.
25. Liu, S., et al. (2013). A supervised multiview spectral embedding method for neuroimaging classification. *The 20th IEEE International Conference on Image Processing (ICIP)* (pp. 601–605). IEEE.
26. Liu, S., et al. (2013). Localized sparse code gradient in Alzheimer's disease staging. In *The 35th Annual International Conference of the IEEE Engineering in Medicine and Biology Society (EMBC)* (pp. 5398–5401). IEEE.
27. Liu, S., et al. (2013). Multifold Bayesian kernelization in Alzheimer's diagnosis. In K. Mori, I. Sakuma, Y. Sato, C. Barillot, & N. Navab (Eds.), *The 16th International Conference on Medical Image Computing and Computer-Assisted Intervention (MICCAI)* (Vol. 8150, pp. 303–310). Berlin: Springer.
28. Liu, S., et al. (2014). Multi-channel neurodegenerative pattern analysis and its application in Alzheimer's disease characterization. *Computerized Medical Imaging and Graphics, 38*, 436–444. ISSN: 0895-6111.
29. Liu, S. Q., et al. (2015). Multi-modal neuroimaging feature learning for multi-class diagnosis of Alzheimer's disease. *IEEE Transactions on Biomedical Engineering, 62*, 1132–1140.
30. Liu, S., et al. (2016). Cross-view neuroimage pattern analysis for Alzheimer's disease staging. *Frontiers in Aging Neuroscience.*
31. Mazziotta, J., et al. (2001). A Probabilistic Atlas and reference system for the human brain: international consortium for brain mapping (ICBM). *Philosophical Transactions of the Royal Society of London. Series B: Biological Sciences, 356*, 1293–1322.
32. Minoshima, S., Frey, K. A., Koeppe, R. A., Foster, N. L., & Kuhl, D. E. (1995). A diagnostic approach in Alzheimer's disease using three-dimensional stereotactic surface projections of Fluorine-18-FDG PET. *Journal of Nuclear Medicine, 36*, 1238–1248.
33. Ng, A. Y., Jordan, M. J., & Weiss, Y. (2002). On spectral clustering: Analysis and an algorithm. *Advances in Neural Information Processing Systems, 2*, 849–856.
34. Park, H. (2012). ISOMAP induced manifold embedding and its application to Alzheimer's disease and mild cognitive impairment. *Neuroscience Letters, 513*, 141–145. ISSN: 0304-3940.
35. Risacher, S. L., et al. (2009). Baseline MRI predictors of conversion from MCI to probable AD in the ADNI cohort. *Current Alzheimer's Research, 6*, 347–361. ISSN: 1875-5828.
36. Schaer, M., et al. (2008). A surface-based approach to quantify local cortical gyrification. *IEEE Transactions on Medical Imaging, 27*, 161–170.
37. Singh, N., Wang, A., Sankaranarayanan, P., Fletcher, P., & Joshi, S. (2012). Genetic, structural and functional imaging biomarkers for early detection of conversion from MCI to AD. In N. Ayache, H. Delingette, P. Golland, & K. Mori (Eds.), *Medical Image Computing and Computer-Assisted Intervention (MICCAI)* (Vol. 7510, pp. 132–140). Berlin: Springer. ISBN: 978-3-642-33414-6.

38. Sokoloff, L., Reivich, M., Kennedy, C., Des-Rosiers, M., et al. (1977). The [14C]Deoxy-Glucose method for the measurement of local cerebral glucose utilization: Theory, procedure and normal values in the conscious and anesthetized albino rat. *Journal of Neurochemistry, 28*, 897–916.
39. Suk, H.-I., Lee, S., & Shen, D. (2013). Latent feature representation with stacked auto-encoder for AD/MCI diagnosis. *Brain Structure and Function, 220*, 841–959.
40. Wachinger, C., Reuter, M., & ADNI, (2016). Domain adaptation for Alzheimer's disease diagnostics. *NeuroImage, 139*, 470–479.
41. Zhang, D., Wang, Y., Zhou, L., Yuan, H., & Shen, D. (2011). Multimodal classification of Alzheimer's disease and mild cognitive impairment. *NeuroImage, 55*, 856–867. ISSN: 1053-8119.
42. Zhang, F., et al. (2015). Pairwise latent semantic asociation for similarity computation in medical imaging. *IEEE Transactions on Biomedical Engineering, 1*, 1–1.
43. Zhu, X., Suk, H.-I., & Shen, D. (2014). A novel matrix-similarity based loss function for joint regression and classification in AD diagnosis. *NeuroImage, 100*, 91–105.

Chapter 6
Alzheimer's Disease Staging and Prediction

Alzheimer's disease (AD), one of the most common and disabling neurodegenerative disorders among aging people, accounts for nearly 70% of all dementia cases.[1]

In the early stage of AD, patients may feel having less energy and spontaneity, then experiencing noticeable decline in memory, language and other cognitive abilities, and becoming depressed. Patients with these signs of neurodegeneration are usually diagnosed as the mild cognitive impairment (MCI), which does not notably interfere with daily activities, but leads to a higher risk of progressing to AD or other forms of neurodegenerative disorders, such as frontotemporal dementia (FTD) and vascular dementia (VD) [5, 7, 21]. So far there is no cure for AD, and medical interventions may only slow down or halt the progression of the disease. Therefore, accurate staging of AD and MCI, especially in the early stage, could assist the neurologists to identify the subjects who are at higher risk and allow them to receive treatments before the onset of dementia symptoms and brain damages.

Most studies on AD staging have a similar work-flow. The primary features are usually extracted from the MRI and/or PET data, and sometimes combined with non-imaging data, such as CSF measures, genetic information and clinical assessments. These biomarkers have been discussed in details in Chaps. 2 and 4. The features are then fed into the various classification models, which are trained on the existing labeled cases and for used to classify future cases. A challenge of the previous studies is how to encode the domain knowledge into the methodology design, since it is essential in AD staging, especially when the datasets are not sufficiently large to train the models. Another challenge is that once the classification models are trained, its performance is bound by the training set. The updates on the training set will not improve the model's performance unless the model is re-trained. This re-training process is a serious bottleneck in AD staging, which impedes the translational impact of the models.

---

[1]Some content of this chapter has been reproduced with permission from [13, 16].

© Springer Nature Singapore Pte Ltd. 2017
S. Liu, *Multimodal Neuroimaging Computing for the Characterization of Neurodegenerative Disorders*, Springer Theses, DOI 10.1007/978-981-10-3533-3_6

This chapter presents two extended workflows to address these challenges. Section 6.1 introduced a domain knowledge-encoded AD staging approach, which aims to optimize the distinction between patient groups by minimizing the intra-group distance and maximizing the inter-group margin [13]. Section 6.2 presents a novel classifier-independent AD prediction model which infers the patient's diagnosis by synthesizing the output probabilities of multimodal biomarkers with no need to re-train any classification models [16].

6.1 Optimized Graph Construction

Domain knowledge, which is usually ignored in classification models, can potentially improve the classification models. In the case of AD staging, for example, the work-flow of previous studies can be refined by including the establish pathology knowledge into the model, such as the empirical conversion rates of MCI and NC subjects. To encode such information, we proposed a novel graph-based method for AD staging. This algorithm features a new global-cost function, which encodes the conversion rates of normal control (NC) subjects and MCI subjects. Global-cost function aligns with the data-cost and smooth-cost functions in graph cut algorithm, and they together ensure the optimal performance of the graph model. We also designed a performance metric to select the most discriminant brain regions/features based on their performance gain. The rest of this section will elaborate on the methodology design of the optimized graph construction in AD staging.

6.1.1 Feature Extraction and Selection

T-tests were carried out on each pre-defined region of interest (ROI) in the brain and the *p-value* of each ROI was used to evaluate its statistical power in distinguishing AD from NC. Notice that, in this context, the *p-values* were not used for the testing any hypotheses, but assessing the statistical power of the ROIs. To select the ROIs/features, we can compare their *p-values* to a significant threshold, usually 0.05 or 0.01, and then select the ROIs/features with *p-values* smaller than the threshold. In this analysis, we took a further step to select the ROIs/features by calculating their performance gain response (PGR), as defined in Eq. 6.1:

$$PGR(i) = 1 - \exp\left(-\frac{(Pr(i) - Pr(i-1))^2}{2\delta^2}\right) \qquad (6.1)$$

where $PGR(i)$ is the PGR of the ith feature in ascending order based on its *p-value*. Features with a PGR greater than the smallest precision unit $\delta = 0.1\%$ were selected.

6.1.2 Graph Construction

Graph cut is a classic technique with a wide range of applications in image segmentation, motion estimation, image restoration, stereo matching and object clustering [4]. A standard graph cut model is usually defined as follows:

$$
E(f|G) = \overbrace{\left[\sum_{v \in V} D_v(f_v)\right]}^{\text{Data Cost}} + \overbrace{\left[\sum_{uv \in E} S_{uv}(f_v, f_v)\right]}^{\text{Smooth Cost}} \tag{6.2}
$$

where f is the label function on the graph $G = \langle V, E \rangle$, V is the vertex set and E is the edge set, *Data Cost* encodes the point's property per se, and *Smooth Cost* regularizes the immediate neighbors to encourage similar labels. Therefore, the image matching or object clustering problems can be solved by minimizing the overall cost, i.e.:

$$
\widehat{f} = \arg\min_{f} E(f|G) \tag{6.3}
$$

We take the neuroimaging feature space as a n-dimensional grid space, $\mathbb{R}$, where subjects of different groups were distributed as samples. In this analysis, the data-cost function encodes the status of each subject depended on the subject's features, and the smooth-cost function encodes the affinity between the subject and its neighbors. Therefore our AD staging problem, which is equivalent to the multi-class classification problem, could be solved by minimizing these cost functions. Assuming the distributions of AD, cMCI, ncMCI and NC subjects were all multivariate Gaussians in $\mathbb{R}$, then we computed each Gaussian's mean (μ_*) and covariance matrix (σ_*) using the expectation maximization algorithm and encoded each subject' negative log likelihood in the Gaussian as the data-cost:

$$
D_v(f_v) = -\ln P(f_v|\mu_*, \sigma_*) \tag{6.4}
$$

where * indicate the group of the subject, and $P(f_v|\mu_*, \sigma_*)$ was the multivariate Gaussian function parameterized by μ_* and σ_*. We then modeled the smooth-cost as follows:

$$
S_{uv}(f_u, f_v) \propto \exp\left(-\frac{\phi(f_u, f_v)^2}{2\tau^2}\right) \cdot \frac{1}{J(u, v)} \tag{6.5}
$$

where $\phi(f_u, f_v)$ is the normalized mutual information of f_u and f_v, τ is the standard deviation of all edges in E, and $J(u, v)$ is the normalized Euclidean distance of u and v in the feature space. This smooth-cost function heavily penalizes the subjects, whose mutual information difference is less than τ, if they are assigned to different clusters; and meanwhile penalizes subjects less serious if they are far away from each other. We evaluated the 5 nearest neighbors only in this analysis.

6.1.3 Domain Knowledge-Based Graph Optimization

It is evident that NC and MCI subjects have different conversion rates to AD. Interestingly enough, within the MCI group, some patients tend to convert to AD quickly, usually in three years, but others remain stable for a long term. We therefore dichotomized MCI group as quick MCI converter (cMCI) and stable MCI non-converter (ncMCI). To encode such prior knowledge, we introduced a new global cost in addition to the cost functions in the original graph cut model, as follows:

$$H(\ell_*) \propto -\ln(\omega(\ell_*)^2) \cdot \frac{1}{\ln(2\pi e)^n |\sigma_*|} \tag{6.6}$$

where $\omega(\ell_*)$ is the empirical conversion rate for subjects with label $\ell_* \in \mathbb{L}$, $\mathbb{L} :=$ $\{\ell_{AD}, \ell_{cMCI}, \ell_{ncMCI}, \ell_{NC}\}$, $\frac{1}{\ln(2\pi e)^n |\sigma_*|}$ is the entropy of $P(u_*, \sigma_*)$, and $|\sigma_*|$ is the determinant of the covariance matrix σ_*. The cMCI's and NC's $\omega(\ell_*)$ were based on their annual conversion rates to AD, and set to 15 and 1.6% respectively [8]. The ncMCI's and AD's $\omega(\ell_*)$ were set to 0.1% to mimic long term immutability. We further normalized the cost function for individual groups using their entropy values. The MCI groups were believed to have greater entropy than the AD and NC groups, since they contained both cMCI and ncMCI. This function penalized less if a MCI subject was mislabeled as AD or NC, whereas penalized more if an AD or NC subject was mislabeled as MCI. Therefore, our AD staging was defined as a cost minimization problem, as in Eq. 6.7. We solved this problem using the α expansion algorithm in the GCO_3.0 library (http://vision.csd.uwo.ca/code/).

$$\widehat{f} = \arg\min_f \left(E(f|G) + \sum_{\ell \in \mathbb{L}} H(\ell|G) \right) \tag{6.7}$$

6.1.4 Performance Evaluation

During feature selection, the feature's PGR was computed as the SVM classification rate. The same features were used for evaluating both the optimized and the original graph using 10-fold cross validation. Notice that these graph models did not require any training datasets to classify the subjects, but assign the subjects into different clusters. The label information was only used to validate the clustering effectiveness. During each iteration of the 10 trials, 90% subjects were evaluated in the proposed models.

Figure 6.1 shows the PGR of individual features in predicting the four diagnostic groups. The features' PGR were closely related to their *p-values*, although several features with significant *p-values* seems not contributive to the clustering performance. The union of the selected features in individual groups finally formed a 52-dimensional feature space, which was used in later analysis. Table 6.1 shows the

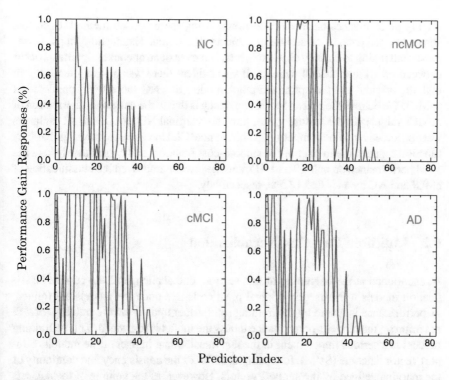

Fig. 6.1 The performance gain responses (%) for individual predictors in AD, cMCI, ncMCI and NC groups. Figure reproduced with permission from [13]

Table 6.1 The classification results of AD, cMCI, ncMCI and NC subjects in the ADNI MRI subset. **True** is the truth labels, **Assign** is the assigned label. Table reproduced with permission from [13]

Algorithm	True\Assign	NC	ncMCI	cMCI	AD
Original graph construction	NC	51.5	48.0	0.5	0
	ncMCI	0.0	99.5	0.5	0
	cMCI	1.3	38.1	60.0	0.0
	AD	2.2	17.2	1.1	79.4
Optimized graph construction	NC	66.2	32.8	0.1	0.0
	ncMCI	0.0	99.1	0.9	0.0
	cMCI	2.5	19.4	72.5	1.9
	AD	3.3	10.6	8.3	77.8

staging performances on these four groups using the proposed method. Each row shows the percentages of assigning subjects to different diagnostic groups. It was most challenging to classify cMCI subjects, and most of erroneous assignments occur between cMCI and ncMCI patients. It was evident there was a strong tendency in both the original and the optimized graph models that NC subject was predicted as ncMCI. We believe the reason for this tendency is due to the extensive distribution of ncMCI subjects in the feature space, thus the marginal NC, cMCI and AD subjects were enforced by smooth function to shift to ncMCI. However, the novel global-cost function reduced the impact of the smooth-cost function, and consequently kept the good performance on ncMCI and AD, and meanwhile improved the classification of cMCI and NC by 14.7 and 12.5%, respectively.

6.2 Multifold Bayesian Kernelization

As mentioned at the beginning of this section, one challenge of the current classification models to be used in clinical practice is that once the models are trained, its performance is bound by the training set. Further updates on the training set will not improve the model's performance unless the model is re-trained. This re-training process is a serious impediment in the fast translational impact of the models. Support vector machine (SVM), for example, enforces the consistency and continuity of the margins defined by the support vectors. However, as the volume of the datasets increase, these support vectors may be no longer optimal for the model. Another challenge is how to combine the multimodal data. Previous studies usually select a number of brain regions/features based on the assumption that other brain regions/features are not important in diagnosis and therefore could be discarded. However, the grouping effects of the features are usually ignored in feature selection. There are also studies that attempted to embed the multimodal features into a unified feature space, such as Partial Least Squares (PLS) [24], ISOMAP [22], multi-view spectral embedding (MSE) [3, 14], yet the existing embedding algorithms seem not able to sufficiently smooth the embedding of multimodal features.

We therefore presented a novel multimodal prediction approach, the Multifold Bayesian Kernelization (MBK), to jointly analyze the multimodal features in the characterization of AD and MCI. MBK first constructs a set of non-linear kernels to obtain the probable diagnosis based on individual features, then deduce the weights of the feature-specific kernels by minimizing the cost of diagnostic errors and kernelization encoding errors using a Bayesian framework, and finally infers the subject's diagnosis by synthesizing the output diagnosis of individual features. MBK is inherently a multi-class model, which is superior to other multimodal methods with two-class models [23, 25, 26].

6.2.1   Algorithm Overview

MBK aims to construct a set of kernels for multimodal features and optimally integrate the probable diagnoses of individual features to enhance the AD prediction. Assume there is a feature vector set X of N subjects with a collection of B multimodal features, M. The subjects' labels are represented as $Y = \{y_1, \ldots, y_N\}$, and the feature for the ith feature, $M^{(i)}$, is represented as $X^{(i)} = \{x_1^{(i)}, \ldots, x_N^{(i)}\} \in \mathbb{R}^{V^{(i)} \times N}$, where $V^{(i)}$ is the dimension of the features. $\mathbf{K}$-step aims to learn a kernel, $K^{(i)}$, for each feature to encode $X^{(i)}$ in such a way that maximizes the preservation of local information. Then in $\mathbf{B}$-step, the contribution of each kernel is evaluated using the Bayesian framework by iteratively minimizing two types of errors: the overall diagnostic errors and the the sum of individual kernels' encoding errors. Finally, in $\mathbf{M}$-step, MBK infers the diagnosis of an unknown subject, $\widetilde{x}$, by synthesizing the probable diagnoses of individual features available to $\widetilde{x}$. Notice that the MBK could ac commodate arbitrary features to make a prediction, even though they do match the existing feature set. Figure 6.2 illustrates the MBK's work-flow.

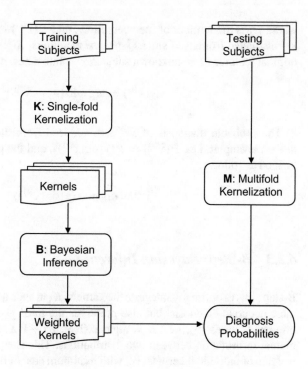

Fig. 6.2 The work-flow of MBK algorithm with three major steps. Figure reproduced with permission from [16]

6.2.2 K-Step: Single-Fold Kernelization

K-step, single-fold kernelization, aims to maximize the preservation of local geometry information and provide a way to infer the subject's label based on its affinity to its labeled neighbors. Such local information is essential in diagnosis of AD since the neuroimaging features usually have high noise to signal ratio and the subjects may not be linearly separable in the feature space, which violates the assumption of many parameterized models, such as SVM and Gaussian Discriminant Analysis (GDA). We constructed the kernel for individual feature using codebook quantization [9]. The affinity propagation algorithm [6] was first used to select a set of exemplars with least square errors to represent the current feature dimension. The kernel, $K^{(i)}$, was then defined as the kernelization codebook of the derived T exemplars, i.e., $K^{(i)} = \{\varepsilon_t\}_{t=1}^T$. Each exemplar, ε_t, was representing a cluster, C_t, in the feature space. Given a ε_t, its marginal distribution of labels was defined as follows:

$$P(y|\varepsilon_t) = \frac{1}{N_t} \sum_{x^{(i)} \in C_t} P(x^{(i)}) \tag{6.8}$$

where N_t is the number of members in C_t, and $P(x^{(i)})$ is the label distribution for $x^{(i)}$ estimated from itself and its k nearest neighbors. $K^{(i)}$ was used to transcribe the original features of an unknown subject, $\widetilde{x}^{(i)}$, into a new codeword as follows:

$$\text{sig}(\widetilde{x}^{(i)}) = \arg\min_{\varepsilon_t} ||\varepsilon_t - \widetilde{x}^{(i)}||^2 \tag{6.9}$$

The probable diagnosis of $\widetilde{x}^{(i)}$ was deduced from the label distribution of its nearest exemplar, i.e., $P(\widetilde{x}^{(i)}) = P(y|\text{sig}(\widetilde{x}^{(i)}))$, and the predicted label of $\widetilde{x}^{(i)}$ was defined as follows:

$$\widehat{y}^{(i)} = \arg\max_y (y|\text{sig}(\widetilde{x}^{(i)})) \tag{6.10}$$

6.2.3 B-Step: Bayesian Inference

B-step aims to optimally integrate the kernels, K, in such a way that not only achieves more accurate diagnosis, but also preserves the local geometry information of the original features [9], i.e., $K = \arg\max(I(K, Y) + I(X, K))$, where $I(*, *)$ is the mutual information between two distributions. This is equivalent to computing the weights of individual kernels, W, with minimum cost of the two types of errors, i.e., the overall diagnostic cost and the sum of cost of kernelization encoding errors, as in Eq. 6.11:

$$\arg\min_W \left[\underbrace{\frac{1}{N} \sum_{j=1}^{N} (\frac{1}{2}||\widehat{y}_{j,M,W} - y_j||^2)}_{\text{Cost of Diagnostic Errors}} \right] + \beta \left[\underbrace{\sum_{i=1}^{M} W(i) \sum_{j=1}^{N} D(P(x_j^{(i)})|P(y|\text{sig}(x_j^{(i)})))}_{\text{Cost of Kernelization Errors}} \right]$$

$$\text{subject to } \sum_{(i=1)}^{M} W(i)=1$$

$$(6.11)$$

where $\widehat{y}_{j,M,W}$ is the synthesized diagnosis using all the features, $D(*,*)$ is the Kullback-Leibler (KL) divergence, and β is the trade-off parameter between these two types of errors. The weights in W were equally initialized, assuming the features are equally important. Then we iteratively updated W: (1) recalculate the cost derived from each kernel after each iteration and then normalize the costs by the total cost as the inferred posterior weights, W'; (2) subtract the average weights of all kernels from W' to derive the change rates of the kernels, dW, (3) use $(W - dW)$ as the new input to the Bayesian framework; (4) repeat this process until the cost can no longer reduced.

6.2.4 M-Step: Multifold Synthesis

M-step aims to infer the probable diagnosis of a given testing subject with a set of features, $\widetilde{M}$. The subject' features were first encoded into the codewords with the single-fold kernels of $\widetilde{M}$ to derive the probable diagnoses based on individual features. The diagnosis probability distribution was then computed from the probable diagnosis of individual kernels, as follows:

$$P(y|\widetilde{x}, \widetilde{M}, W) = \sum_{i:\{M^{(i)} \in \widetilde{M}\}} W(i) P(y|\text{sig}(\widetilde{x}^{(i)})) \qquad (6.12)$$

where $\text{sig}(\widetilde{x}^{(i)})$ is the codeword of $\widetilde{x}$ deduced from the ith kernel. The synthesized diagnosis of $\widetilde{x}$ was defined as follows:

$$\widehat{y}_{j,\widetilde{M},W} = \arg\max_y P(y|\widetilde{x}, \widetilde{M}, W) \qquad (6.13)$$

Notice that the dimensionality of features in $\widetilde{M}$ does not need to match M, since the outputs of the M-step are the diagnostic probability distributions, which can be straightforwardly merged to make the diagnosis. Meanwhile it does not require re-training the model, although more features may lead to more deterministic diagnoses. Such flexibility makes MBK a more practical model than the parameterized models.

6.2.5 Performance Evaluation

The proposed MBK algorithm was validated on the ADNI MRI-PET subset. More details about this dataset can be found in Chap. 3, Sect. 3.3. Four types of features were extracted from the multimodal neuroimaging data, including the mean index (M-IDX) from PET [1, 10, 11, 17, 18], and the grey matter volume (GMV) [12–14], solidity (SLD) and convexity (CNV) [15, 19, 20] features from MRI. These features were discussed in Chap. 4. We compared MBK to three state-of-the-art neuroimaging classification models, including (1) ISOMAP [22], which was used as the benchmark of the feature embedding algorithms; (2) elastic net (EN) [23], which was used as the benchmark of the feature selection algorithms; (3) domain-knowledge-learning graph cut (DKL-GC) algorithm [12], which was used as the benchmark of supervised learning algorithms. More specifically, a global-cost function was designed to encode the different AD conversion rates and minimize the erroneous classification of cMCI. DKL-GC was solved by the GCO_3.0 library [4]. The features processed by EN and ISOMAP were fed into the SVM with Gaussian kernels, which was implemented using the LIBSVM library [2]. We used the grid-search method to find the optimal trade-off parameter (C) and the Gaussian parameter (γ) in SVM, and the cost function weight parameters in DKL-GC. The parameter settings of MBK were set by pilot experiments, i.e., $[k, \beta] = [5, 0.5]$ in this analysis.

Five-fold cross-validation paradigm was adopted by all the methods in performance evaluation with a 20% subset of the entire dataset used as the testing set and the rest subset used as training set in each iteration. Notice that in the MBK experiments, the same training set was used to construct the single-fold kernels in **K**-step as well as to derive the kernel parameters in **B**-step in each iteration. The average classification rates of four diagnosis groups were used to evaluate the performance of these methods. In another experiment, we dichotomized the multimodal feature set based on their modalities, including 83 PET features, and 249 MRI features. Then the predictions in **B**-step were made using the PET features, MRI features and the merged PET and MRI features, respectively. Figures 6.3 and 6.4 illustrate the average diagnosis accuracy and the cost of errors in each iteration while updating the kernels' weights. The error bars indicate the means and standard deviations of the 5 trials in cross-validation. It was evident that the merged PET and MRI features reached the highest accuracy and lowest error cost after 11 iterations and its performance leveled off after 15 iterations.

Table 6.2 shows the results of MBK, ISOMAP and EN with SVMs, and DKL-GC. MBK outperformed the other classification models in all of the groups, with a dramatically high average accuracy of 74.2% compared to 38.4% of the ISOMAP, 54.3% of EN, and 63.29% of DKL-GC. The ISOMAP method had the worst performance, which suggested that it was not suitable for multimodal feature embedding. EN preserved the correlation between features by adding the l_1 and l_2 regulations on the feature variables, which encouraged the grouping effect and led to better results than ISOMAP. The DKL-GC algorithm purposed to enhance the prediction of cMCI by increasing the penalty for a erroneous cMCI prediction, as a result it achieved a

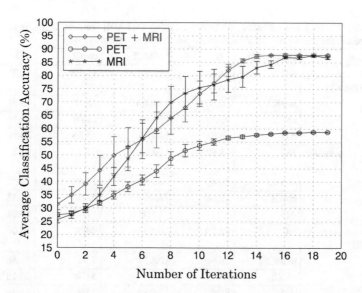

Fig. 6.3 The average classification accuracy of the training set in each iteration of **B**-Step in MBK. Figure reproduced with permission from [16]

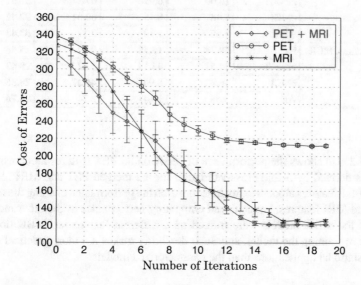

Fig. 6.4 The cost of diagnostic errors and kernelization errors outputted in each iteration of **B**-Step in MBK. Figure reproduced with permission from [16]

markedly higher cMCI classification rate than ISOMAP and EN, whereas the classification of ncMCI was compromised due to such penalty function design. The MBK algorithm, on the other hand, was inherently a multi-class model and relied on no domain knowledge, which might bias the performance on either patient group.

Table 6.2 The diagnosis accuracy (%) evaluated using PET-MRI features. **Dgns.** is the ground truth, **Prdt.** is the prediction. Table reproduced with permission from [16]

Algorithm	Dgns. \Prdt.	CN	ncMCI	cMCI	AD
Feature embedding: ISOMAP-SVM	NC	**34.33**	38.80	15.60	11.27
	ncMCI	26.64	**38.86**	15.12	19.38
	cMCI	20.30	34.46	**21.08**	24.16
	AD	16.81	25.66	18.56	**38.96**
Feature selection: EN-SVM	NC	**60.57**	29.13	4.13	6.17
	ncMCI	27.43	**43.56**	11.69	17.32
	cMCI	17.96	33.64	**25.06**	23.33
	AD	5.71	19.05	11.43	**63.81**
Supervised learning: DKL-GC	NC	**64.29**	0.00	0.65	35.06
	ncMCI	26.96	**38.24**	2.94	31.86
	cMCI	21.64	6.72	**51.49**	20.15
	AD	8.24	7.06	2.94	**81.76**
The proposed: **MBK**	NC	**86.00**	6.50	1.00	6.50
	ncMCI	10.00	**66.96**	0.43	22.61
	cMCI	8.48	8.48	**60.61**	22.42
	AD	5.65	8.70	2.17	**83.48**
PET features MBK	NC	**59.74**	15.58	9.09	15.58
	ncMCI	24.51	**43.14**	3.92	28.43
	cMCI	16.42	8.96	**46.27**	28.36
	AD	3.53	16.47	8.24	**71.76**

Table 6.2 also shows the performance of MBK on 83 PET features alone using the weights derived by 5-fold cross-validation for the merged PET and MRI features. Using PET features alone was not able to match the performance of using the merged PET and MRI features, but achieved competing performance as the other methods, even if the MBK models were trained on a different feature set. MBK does not require re-training the model, and Such flexibility makes it a more practical choice in translational applications than the parameterized models.

6.3 Summary

This chapter presents two methods to investigate the entire spectrum of AD, i.e., NC, ncMCI, cMCI and AD, for the disease staging and prediction. AD staging aims to find the boundaries between patients at different stages as the disease progresses by separating them into groups with the minimum intra-group variance and the maximum inter-group margin. The goal of AD prediction is to infer the patient's status, i.e., AD or MCI, by comparing to the other patients with confirmed diagnoses. Domain knowledge can be used to enhance the staging of AD, especially when there are limited datasets available for training. A new global-cost function that encodes the empirical AD conversion rates for different diagnostic groups was proposed and integrated to the classic graph cut algorithm. This global-cost function lessened the negative impact of the smooth-cost function in the original algorithm and demonstrates potential for distinction of different stages of AD. However, such the domain knowledge might lead to biased classification.

In AD prediction/classification, the classification accuracy is always constrained by the training set for the classifiers, e.g., support vector machine (SVM) enforces the global consistency and continuity of the boundaries by identifying the support vectors. However, it requires re-training to transfer the information of any new datasets to the classifier. To overcome this limitation, a novel diagnosis algorithm for AD prediction is proposed. It differs from the parameterized classification models in that: (1) it models the diagnosis process as a synthesis analysis of multimodal biomarkers; (2) it adopts a novel diagnosis scheme synthesizing the outputted diagnosis probabilities of individual features instead of combining the inputted features as a whole. The preliminary results showed that the MBK algorithm outperformed the state-of-the-art classification models and has a great potential in computer aided diagnosis of AD and MCI.

References

1. Cai, W., et al. (2010). 3D neurological image retrieval with localized pathology-centric CMR-Glc patterns. In *The 17th IEEE International Conference on Image Processing (ICIP), IEEE* (pp. 3201–3204).
2. Chang, C. C., & Lin, C. J. (2011). LIBSVM: A library for support vector machines. *ACM Transactions on Intelligent Systems Technology (ACM TIST), 2*, 27:1–27:27. ISSN: 2157-6904.
3. Che, H., et al. (2014). Co-neighbor multi-view spectral embedding for medical content based retrieval. In *IEEE International Symposium on Biomedical Imaging: From Nano to Macro (ISBI), IEEE* (pp. 911–914).
4. Delong, A., Osokin, A., Isack, H., & Boykov, Y. (2012). Fast approximate energy minimization with label costs. *International Journal of Computer Vision (IJCV), 96*, 1–27. ISSN: 0920-5691.
5. Dubois, B., & Albert, M. (2004). Amnestic MCI or prodromal Alzheimer's disease? *Lacent Neuology, 3*, 246–248.
6. Frey, B. J., & Dueck, D. (2007). Clustering by passing messages between data points. *Science, 315*, 972–976. ISSN: 1095-9203.
7. Jicha, G., Parisi, J., Dickson, D., Johnson, K., et al. (2006). Neuropathologic outcome of mild cognitive impairment following progression to clinical dementia. *Archives of Neurology, 63*, 674–681.
8. Kalaria, R. N., et al. (2008). Alzheimer's disease and vascular dementia in developing countries: Prevalence, management, and risk factors. *The Lancet Neurology, 7*, 812–826. ISSN: 1474-4422.
9. Lazebnik, S., & Raginsky, M. (2009). Supervised learning of quantizer codebooks by information loss minimization. *IEEE Transactions on Pattern Analysis and Machine Intelligence (PAMI), 31*, 1294–1309. ISSN: 0162-8828.

10. Liu, S., et al. (2011). Generalized regional disorder-sensitive-weighting scheme for 3D neuroimaging retrieval. In *The 33rd Annual International Conference of the IEEE Engineering in Medicine and Biology Society (EMBC), IEEE* (pp. 7009–7012).
11. Liu, S., Cai, W., Wen, L., & Feng, D. (2012). Semantic-word-based image retrieval for neurodegenerative disorders. *Journal of Nuclear Medicine, 53*, 2309.
12. Liu, S., Cai, W., Wen, L., & Feng, D. (2013). Multi-channel brain atrophy pattern analysis in neuroimaging retrieval. In *IEEE International Symposium on Biomedical Imaging: From Nano to Macro (ISBI), IEEE* (pp. 206–209).
13. Liu, S., Cai, W., Wen, L., & Feng, D. (2013). Neuroimaging biomarker based prediction of Alzheimer's disease severity with optimized graph construction. In *IEEE International Symposium on Biomedical Imaging: From Nano to Macro (ISBI), IEEE* (pp. 1324–1327).
14. Liu, S., et al. (2013). A supervised multiview spectral embedding method for neuroimaging classification. In *The 20th IEEE International Conference on Image Processing (ICIP), IEEE* (pp. 601–605).
15. Liu, S., et al. (2013). Localized sparse code gradient in Alzheimer's disease staging. In *The 35th Annual International Conference of the IEEE Engineering in Medicine and Biology Society (EMBC), IEEE* (pp. 5398–5401).
16. Liu, S., et al. (2013). Multifold Bayesian kernelization in Alzheimer's diagnosis. In K. Mori, I. Sakuma, Y. Sato, C. Barillot, & N. Navab (Eds.), *The 16th International Conference on Medical Image Computing and Computer-Assisted Intervention (MICCAI)* (Vol. 8150, pp. 303–310). Berlin Heidelberg: Springer.
17. Liu, S., et al. (2014). Multi-channel neurodegenerative pattern analysis and its application in Alzheimer's disease characterization. *Computerized Medical Imaging and Graphics, 38*, 436–444. ISSN: 0895-6111.
18. Liu, S. Q., et al. (2014). High-level feature based PET image retrieval with deep learning architecture. *Journal of Nuclear Medicine, 55*, 2018.
19. Liu, S., et al. (2015). Subject-centered multi-view neuroimaging analysis. In *The 22nd IEEE international conference on image processing (ICIP)*. IEEE.
20. Liu, S., et al. (2016). Cross-view neuroimage pattern analysis for Alzheimer's disease staging. *Frontiers in Aging Neuroscience*.
21. Nettiksimmons, J., DeCarli, C., Landau, S., & Beckett, L. (2014). Biological heterogeneity in ADNI amnestic mild cognitive impairment. *Alzheimer's & Dementia,10*, 511–521.
22. Park, H. (2012). ISOMAP induced manifold embedding and its application to Alzheimer's disease and mild cognitive impairment. *Neuroscience Letters, 513*, 141–145. ISSN: 0304-3940.
23. Shen, L., et al. (2011). Identifying neuroimaging and proteomic biomarkers for MCI and AD via the elastic net. In T. Liu, D. Shen, L. Ibanez, & X. Tao (Eds.), *Multimodal Brain Image Analysis (MBIA)* (Vol. 7012, pp. 27–34). Berlin Heidelberg: Springer. ISBN: 978-3-642-24445-2.
24. Singh, N., Wang, A., Sankaranarayanan, P., Fletcher, P., & Joshi, S. (2012). Genetic, structural and functional imaging biomarkers for early detection of conversion from MCI to AD. In N. Ayache, H. Delingette, P. Golland, & K. Mori (Eds.), *Medical Image Computing and Computer-Assisted Intervention (MICCAI)* (Vol. 7510, pp. 132–140). Berlin Heidelberg: Springer. ISBN: 978-3-642-33414-6.
25. Ye, J., et al. (2012). Sparse learning and stability selection for predicting MCI to AD conversion using baseline ADNI data. *BMC Neurology, 12*, 46. ISSN: 1471-2377.
26. Zhang, D., Wang, Y., Zhou, L., Yuan, H., & Shen, D. (2011). Multimodal classification of Alzheimer's disease and mild cognitive impairment. *NeuroImage, 55*, 856–867. ISSN: 1053-8119.

Chapter 7
Neuroimaging Content-Based Retrieval

Medical content-based retrieval (MCBR) is progressing rapidly with the advances in database systems, computer vision and medical informatics. MCBR has a wide range of medical applications, including medical imaging data management, clinical training and education. Most importantly, it provides access to the cases of previously diagnosed patients, thus is able to support clinical decisions for future cases [4, 27].[1]

Many MCBR systems have been proposed, mostly for single-modal data, such as High Resolution Computed Tomography (HRCT) [8], Positron Emission Tomography (PET) [3, 5, 6, 9–15, 20], Single Photon Emission Computed Tomography (SPECT) [29], Magnetic Resonance Imaging (MRI) [16, 17, 24, 32, 36], functional MRI (fMRI) [2], diffusion MRI (dMRI) [23], and multimodal neuroimaging [7, 18, 21, 37–39]. More details of these studies can be found in Chap. 2, Sect. 2.4.2.

With increasing attention given to multimodal analysis, a number of naive methods have been proposed to integrate the multimodal features, e.g., concatenating multimodal features into a high-dimensional vector, selecting features in the high-dimensional space [12, 25, 30], embedding the subjects in a new feature space [28], and using a combination of these methods to overcome the 'curse of dimensionality' [16, 22]. However, none of these methods was able to eliminate the redundancy in a plurality of features and maximize the inter-dimension variance at the same time. Recently, multimodal MCBR analysis has refueled by the research efforts on the multi-view embedding (ME) methods, such as Multi-View Spectral Embedding (MSE) [19, 33] and Multi-View Local Linear Embedding (MLLE) [31]. ME methods, based on manifold-learning, enable exploring the geometric structures of local patches across multiple feature spaces and then smoothly aligning them in a unified

---

[1]Some content of this chapter has been reproduced with permission from [21, 23, 26].

© Springer Nature Singapore Pte Ltd. 2017
S. Liu, *Multimodal Neuroimaging Computing for the Characterization of Neurodegenerative Disorders*, Springer Theses, DOI 10.1007/978-981-10-3533-3_7

feature space with maximum geometric information. Although ME methods are able to preserve the inter-dimension variance, the unpredictable variability of the unseen queries pose a bottleneck to the current retrieval systems.

Therefore, we present a novel propagation graph fusion (PGF) algorithm for subject-centered multimodal MCBR in this Chapter. PGF is an unsupervised method, requiring no prior knowledge about the features or the queries, and adaptively reshaping the connections between the subjects for different queries, thus retrieving more relevant subjects. Section 7.1 introduces the basic PGF method and summarizes the retrieval results based on the ADNI MRI-PET subset. Section 7.2 presents an improved PGF method and demonstrates the improvements over original PGF on the ADNI MRI and dMRI subsets.

7.1 Propagation Graph Fusion

7.1.1 Propagation Graph Construction

Assuming we have extracted N_v features from N_d subjects in the database ($N_v = 4$, $N_d = 331$ in this analysis), $x_i^{(n)}$ represents the feature of the ith subject in the nth feature space. To preserve the local geometric information of the subjects across multiple feature spaces, we defined the local neighborhood of each subject in each feature space as follows:

$$X_i^{(n)} = \left\{ x_i^{(n)}, x_{i_1}^{(n)}, \ldots, x_{i_k}^{(n)} \right\} \tag{7.1}$$

where $x_{i_1}^{(n)}, \ldots, x_{i_k}^{(n)}$ are the k nearest neighbors of $x_i^{(n)}$ measured by standard Euclidean distance. We further modeled each subject as a node and then connected them with edges, whose weights were based on the Jaccard coefficients of their neighborhoods, as follows:

$$w(x_i^{(n)}, x_j^{(n)}) = \frac{|X_i^{(n)} \cup X_j^{(n)}|}{|X_i^{(n)} \cap X_j^{(n)}|} \tag{7.2}$$

These edges formed a network, and their weights showed the affinity between the connected subjects. Notice that we used Jaccard coefficient other than Euclidean distance to measure the affinity, since the distance measurement had varying mean and variance across different feature spaces and likely lead to a single dominant feature or a set of less discriminative features.

PGF constructed a network for each query, x_q, and the subjects in the database, centering at x_q. Let x_q be the source, and propagate along the possible paths in the network. The propagation started from x_q to its k nearest neighbors, $x_{q_1}^{(n)}, \ldots, x_{q_k}^{(n)}$, in the nth feature space, and continued until no more nodes could be visited. If there was a path connecting the query to a node, then the relevance of the query and the

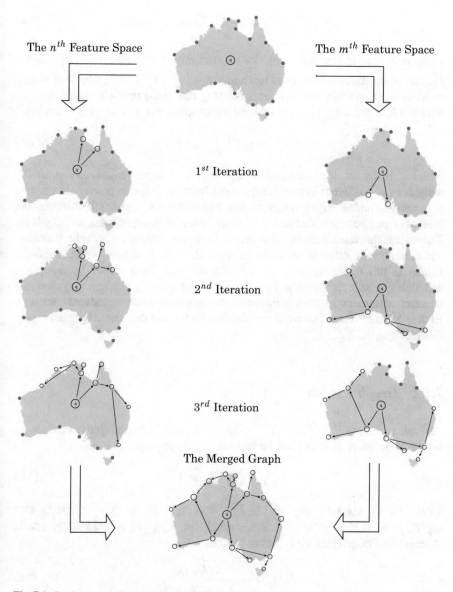

Fig. 7.1 Basic propagation graph construction and fusion

node was proportional to the network flow passing through that node. The network flow was computed as sum of weights over the path, since the weights reflected the affinity of the query and the subjects. We further took into account the damping effect of propagation, and updated the edge weights for x_q as in Eq. 7.3:

$$w_q^{(n)}(x_i^{(n)}, x_j^{(n)}) = \alpha^{t_q(x_i^{(n)}, x_j^{(n)})} \cdot w(x_i^{(n)}, x_j^{(n)}) \tag{7.3}$$

where α is a parameter that regularize the damping effect, and $t_q(x_i^{(n)}, x_j^{(n)})$ is the number of iterations necessary to reach the edge $(x_i^{(n)}, x_j^{(n)})$. If an edge was visited multiple times, we then selected the smallest t_q among the multiple values to maximize the flow capacity. Then the updated paragraph graph was saved as in Eq. 7.4

$$G_q^{(n)} = \left(V_q^{(n)}, E_q^{(n)}, w_q^{(n)}\right) \tag{7.4}$$

where V is the set of visited nodes, E is the set of visited edges, and w are the weights of E. Different feature spaces could result in different propagation graphs with varying nodes, edges, weights, sizes and structures. Figure 7.1 illustrates the process of graph construction in different feature spaces based on 2 nearest neighbors. The propagation starts from the same query to its nearest neighbors in the first iteration and then flows to other nodes alone the possible paths in the following iterations. Circles in this figure represent the subjects that have been visited by the query, and their size reflects the subject's relevance to the query. The edges also show the direction of the network flow. In this example, the propagation stops after 3 iterations. Both m^{th} and n^{th} features have some unvisited nodes, and they result in dramatically different graphs.

7.1.2 Graph Fusion

Multiple graphs, $G_q^{(1)}, G_q^{(2)}, \dots, G_q^{(N_v)}$, can be obtained from multiple feature spaces for the same query, and they can be fused into a single graph, as follows:

$$G_q = \left(V_q, E_q, w_q\right) \tag{7.5}$$

where $V_q = \cup_{N_v} V_q^{(n)}$, $E_q = \cup_{N_v} E_q^{(n)}$, and $w_q(i, j) = \sum_{N_v} w_q^{(n)}(i, j)$ with $w_q^{(n)}(i, j) = 0$ if $(i, j) \notin E_q^{(n)}$. After graph fusion, a $|V| \times |V|$ affinity matrix A_q can be deduced from for G_q, as in Eq. 7.6:

$$A_q(i, j) = \frac{w_q(i, j))}{\sum_{l=|V|} w_q(i, l)} \tag{7.6}$$

Figure 7.1 also shows an example of fusing two graphs derived from different feature spaces for the same query. The size of the circles is proportional to the edge weights of the path connecting the node and the query. In this example, some nodes have been included in both graphs, therefore, their circles become larger in the fused graph and the circles of other nodes remain the same as in single graphs. Consider A as a transition matrix, whose values indicate likelihood of the subjects being visited by the query, then there exists an equilibrium state, i.e., the probabilities over the nodes to be visited by the query become stable. The probability distribution therefore can be used as a good estimate of relevance between the query and the nodes. In this analysis, we used the PageRank algorithm to derive the equilibrium distribution [35].

7.1.3 Validation on the ADNI MRI-PET Subset

The proposed PGF algorithm was validated on the ADNI MRI-PET subset, as described in Chap. 3, Sect. 3.3. We used a query-by-example paradigm and a leave-one-out cross-validation strategy in our experiments. The retrieval results were evaluated using the modified Mean Average Precision (MAP), as follows:

$$MAP = \frac{\sum_{q=1}^{Q} \sum_{k=1}^{K_q} (p_q(k) \cdot rel_q(k))/T_q}{Q} \qquad (7.7)$$

where q is the query index, Q is the number of queries, k is the subject's ranking in sequence of retrieved subjects, K_q is the number of retrieved subjects, $p_q(k)$ is the precision at cut-off k, $rel_q(k)$ is the relevance score of the kth retrieval result, and T_q is the number of truly relevant subjects of the query in the database. Since MCI is always considered as a pre-symptomatic status of AD, conferring a high conversion rate, we used the following relevance criteria to compute the MAPs, as shown in Table 7.1.

We carried out two retrieval experiments on the ADNI MRI-PET subset. The first experiment was conducted on single-modal features, including the mean index (M-IDX) from PET, the lesion-centric features derived from PET by the DoG operator (DoG-C,M,Z), the regional grey matter volume (GMV) from MRI, and the brain atrophy features from MRI captured by convex hull (CVX,SLD). Details of these features could be found in Chap. 4. These single-modal features were used as baselines to verify the effectiveness of multimodal features. The second experiment was conducted on multimodal features using the proposed PGF method, compared to the concatenation method (CONCAT), Elastic Net (EN), and Multi-View Spectral Embedding (MSE). The number of relevant results in MAP was set to 5 for all of the multimodal methods. The used grid-search to determine the parameters of PGF, including the number of nearest neighbors ($k = 4$) and the propagation parameter ($\alpha = 0.1$). The parameters and hyper-parameters for other methods were optimized through random search [1].

Table 7.1 The relevance criteria for AD, MCI, and NC, used in computing the MAP (%). Table reproduced with permission from [21]

	NC	MCI	AD
NC	1	0.25	0
MCI	0.25	1	0.25
AD	0	0.25	1

Table 7.2 The $MAP \pm STD$ (%) in retrievals of NC, MCI and AD subjects using single-modal features, evaluated on the ADNI MRI-PET subset. Table reproduced with permission from [21]

	M-IDX	DoG-C,M,Z	GMV	CVX,SLD
NC	51.9 ± 13.4	53.1 ± 13.9	**53.7 ± 13.0**	51.3 ± 13.8
MCI	75.4 ± 11.7	74.9 ± 12.0	76.9 ± 11.6	**77.8 ± 12.1**
AD	49.0 ± 12.9	51.7 ± 13.2	47.3 ± 11.8	**53.9 ± 13.5**
Average	63.2 ± 13.9	63.9 ± 14.0	63.9 ± 13.7	**65.5 ± 14.2**

Table 7.3 The $MAP \pm STD$ (%) in retrievals of NC, MCI and AD subjects using multimodal features, evaluated on the ADNI MRI-PET subset. Table reproduced with permission from [21]

	CONCAT	EN	MSE	PGF
NC	52.7 ± 13.8	58.1 ± 5.6	59.6 ± 6.4	**73.5 ± 5.5**
MCI	77.4 ± 12.0	82.6 ± 5.1	82.5 ± 6.0	**84.9 ± 4.8**
AD	54.3 ± 13.2	62.7 ± 5.2	64.1 ± 5.6	**68.8 ± 6.0**
Average	65.7 ± 14.1	71.8 ± 12.2	72.4 ± 12.3	**78.1 ± 11.4**

Table 7.2 summarizes the retrieval results obtained from single-modal features. The best performance in each subject group is highlighted in bold face. MRI features, in most cases, worked better than PET features in retrieving the AD and MCI subjects. The Convex features, in particular, achieved the best retrieval result at 77.8% in MCI, 53.9% in AD, and 65.5% on average of the three groups combined. The best performance in NC was obtained by GMV at 53.7%.

Table 7.3 summarizes the retrieval results obtained from multimodal data. The naive CONCAT method worked slightly better than the best single-modal feature. After EN was applied to select the most important features, the performance was further improved by 5.4% in NC retrieval, 5.2% in MCI retrieval, and 8.4% in AD retrieval. The average retrieval improvement was 6.1%. Notice that EN was a supervised method, which required the label information of the subjects to select the feature. MSE, which was an unsupervised method, achieved nearly matched performance with EN in MCI retrieval, and outperformed EN in NC and AD retrieval. PGF had the best performance on all of the groups with an average MAP of 78.1%. It had a marked performance in NC retrieval at 73.5%, which was 20.8% higher than the CONCAT method, 15.4% higher than EN, and 13.9% higher than MSE.

7.2 Geometric Mean Propagation Graph Fusion

The graph fusion method described in Sect. 7.1.2 is equivalent to calculating the arithmetic mean of edge weights in the graphs. It was easy to implement this method, but it might not be sufficient to differentiate some challenging cases due to the offset effect of arithmetic mean, e.g., fusing two moderate-weighted edges might lead to the same result as fusing a high-weighted edge and a low-weighted edge. To minimize such offset effect, we were motivated to replace the arithmetic mean with the geometric mean, which could measure the conformity of the retrieval results across multiple graphs.

7.2.1 Affinity Matrix Construction

The distances between two images in different feature spaces reflect their geometric relationships in different views. To preserve such geometric information, we constructed a neighborhood for each subject in each feature space, same as in Sect. 7.1.1. Assuming N_v features were extracted from N_d subjects in the database $\mathbb{D}$, the feature vector of one subject in the database can be denoted by x_i, and the feature set can be denoted by X. The neighborhood of x_i in the nth feature space was formed by itself and its k nearest neighbors, as follows:

$$X_i^{(n)} = x_i^{(n)}, x_{i_1}^{(n)}, \ldots, x_{i_k}^{(n)} \tag{7.8}$$

We then established the connections between subjects by measuring the consistency of their neighborhoods using the Jaccard coefficient as in Eq. 7.9:

$$w(x_i^{(n)}, x_j^{(n)}) = \frac{|X_i^{(n)} \cap X_j^{(n)}|}{|X_i^{(n)} \cup X_j^{(n)}|} \tag{7.9}$$

These connections formed a directed graph. Given a query x_q, we applied the same procedure to construct its neighborhood $X_q^{(n)}$, and used the query as the origin and let x_q walk down the paths to its k nearest neighbors $x_{q_1}^{(n)}, \ldots, x_{q_k}^{(n)}$. x_q continued to walk from $x_{q_1}^{(n)}, \ldots, x_{q_k}^{(n)}$ to their nearest neighbors in $X_{q_1}^{(n)}, \ldots, X_{q_k}^{(n)}$ in the next iteration, and kept walking until reach the end of all of the paths. The intuition was that the longer it took for the query to reach a subject, the less relevant that subject was. We again took damping effect into consideration, and updated the connection weights with regard to a specific query as in Eq. 7.10:

$$w'(x_i^{(n)}, x_j^{(n)}) = \alpha^{t_q(x_i^{(n)}, x_j^{(n)})} \cdot w(x_i^{(n)}, x_j^{(n)}) \tag{7.10}$$

where α is a weight decay parameter that regularize the damping effect, and $t_q(x_i^{(n)}, x_j^{(n)})$ is the number of iterations necessary to reach the link, $(x_i^{(n)}, x_j^{(n)})$. If a link was visited multiple times, the smallest t_q was chosen for it. The updated weights were saved in a $N_d \times N_d$ affinity matrix, A, as in Eq. 7.11:

$$A(i, j) = w'(x_i^{(n)}, x_j^{(n)}) \tag{7.11}$$

A could be very sparse since there were many edges missed out during the propagation. Sparsity usually simplified the computation, but this was not the case in our algorithm. A sparse *A* would reduce the coverage of the relevant subjects when fusing the affinity matrices using the geometric means (see details in Sect. 7.2.2). Therefore, we applied Laplace smoothing to *A*, which added a small value of $1/N_d$ to all of the elements in *A*. We finally normalized *A* to enforce the row-stochasticity, i.e., each row summed up to 1, as in Eq. 7.12:

$$A'(i, j) = \frac{A(i, j) + 1/N_d}{\sum_{X_j \in \mathbb{D}} A(i, l) + 1} \tag{7.12}$$

Figure 7.2 shows an example of the subject-centered affinity matrix construction. The connections between subjects are based on 2 nearest neighbors, and they form a directed graph, as illustrated in 'Initial Connections'. The query, as indicated by the asterisk at the center of the graph, visits its nearest neighbors in the first iteration and then propagates to other neighbors alone the paths in the following iterations. The hollow circles represent the subjects that have been visited by the query, and the solid lines show the propagation routes. In this example, the propagation stops after 5 iterations, but there is still one unvisited subject and some unweighted edges. Laplace smoothing finally adds a weak weight to all the edges, as indicated by the dashed lines.

7.2.2   Affinity Matrix Fusion

Multiple affinity matrices, $A^{(1)}, \ldots, A^{(N_v)}$, could be obtained from multiple feature spaces. Rather than using the arithmetic mean as in PGF [21], we used the geometric mean to fuse the affinity matrices to eliminate the offset effect of arithmetic mean, as in Eq. 7.13:

$$A^*(i, j) = {}^{N_v}\sqrt{\Pi_{n=1}^{N_v} A^{(n)}(i, j)} \tag{7.13}$$

Figure 7.3 illustrates the multiple graph fusion with the arithmetic mean and the geometric mean, respectively. Assuming there are three subjects in the database marked by different shapes. They all have three types of features, as seen in the three different feature spaces. The concentric circles indicate the degree of similarity between the query and the subjects in each feature space. In this toy example, it is evident that arithmetic mean is sufficient to differentiate the subjects, which results in the same similarity level. However, the geometric mean measures the conformity of the retrieval results across multiple views, thus is able to better differentiate the subjects compared to the arithmetic mean.

One condition of using the geometric mean is that $A^{(n)}(i, j) \neq 0$, otherwise the link $(x_i^{(n)}, x_j^{(n)})$ will be disconnected, even if there exist connections in other feature

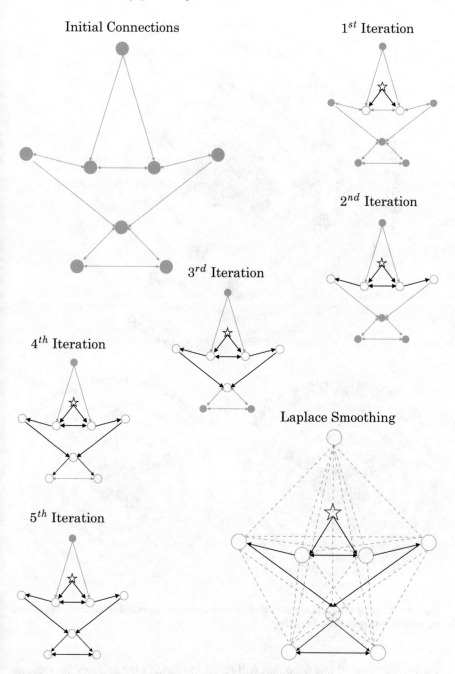

Fig. 7.2 A toy example of the subject-centered affinity matrix construction

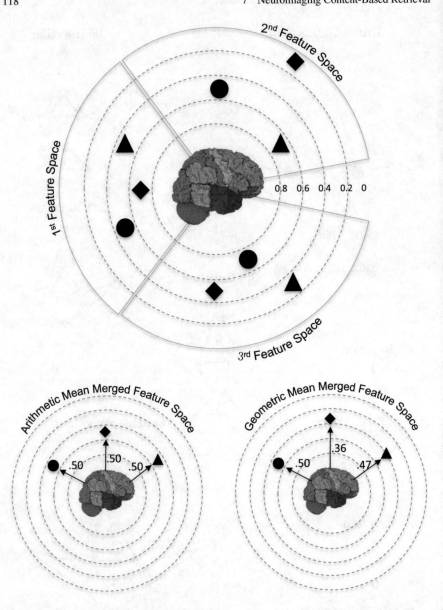

Fig. 7.3 The comparison of the results derived from arithmetic mean and geometric mean. Figure reproduced with permission from [26]

spaces. Violation of this condition will result in blocking the paths to the subjects in the feature space and failure to retrieve the most relevant results. We therefore solved this problem using Laplace smoothing in the affinity matrix construction, as described in Sect. 7.2.1.

The values in A^* reflect the affinity of the subjects to the query. Consider A^* as a transition matrix, there exists an equilibrium state reflecting the probabilities over the subjects to be visited by the query, and the subjects can be re-ranked according to this probability distribution. We applied the PageRank algorithm to obtain the equilibrium state distribution in this analysis.

7.2.3 Validation on the ADNI MRI Subset

The PGF with geometric mean (PGFg) was compared to the PGF with arithmetic mean (PGFa) [21], as well as the naive concatenation method (CONCAT) method, and MSE [33] on the ADNI MRI subset. We used a query-by-example paradigm and the leave-one-out cross-validation strategy in all of the experiments. The performance of these methods was evaluated using the modified MAP with three different cut-off numbers (1, 3, 5) of the relevant results. The number of nearest neighbors for constructing the local neighborhood was set to $k = 10$ for PGF and MSE. Other parameters and hyper-parameters for these methods were optimized via random search [1].

Table 7.4 summarizes the results of different retrieval methods. The CONCAT method, had a poor performance, worse than the using the GMV feature alone, since it was heavily affected by the large variance in the multimodal features. MSE showed a better performance than CONCAT, with slight improvements of 1.8% in MAP(1), 1.8% in MAP(3) and 1.7% in MAP(5). Greater improvements were made by PGFa and PGFg, which both benefited from the adaptive work-flows for the subject-centered analysis. PGFa increased 4.4, 3.6 and 2.8% over MSE in MAP(1), MAP(3) and MAP(5), respectively. The advantage of geometric mean was evident that PGFg further improved PGFa by 3.4% in MAP(1), 1.6% in MAP(3) and 0.9% in MAP(5).

7.2.4 Validation on the ADNI dMRI Subset

We further validated the proposed PGF methods on the ADNI dMRI subset, which contained 233 subjects with 3D axial brain dMRI scans. More details about this

Table 7.4 The performance in MAP (%) of different retrieval methods, evaluated on the ADNI MRI subset. Table reproduced with permission from [26]

	CONCAT	MSE	PGFa	PGFg
MAP(1)	60.8	62.6	67.0	70.4
MAP(3)	53.0	54.8	58.4	60.0
MAP(5)	49.6	51.3	54.1	55.0

Table 7.5 The MAP (%) of the naive concatenation method (CONCAT) and the PGF algorithm with and without EN feature selection (EN-*), evaluated on the ADNI dMRI subset. Table reproduced with permission from [23]

Method	NC	MCI	AD	Average
CONCAT	34.04	56.39	74.26	54.90
PGFg	24.90	52.07	56.51	44.49
EN-CONCAT	39.84	55.66	74.77	56.76
EN-PGFg	72.57	52.27	74.54	66.46

dataset can be found in Chap. 3, Sect. 3.4. We again used the leave-one-out cross-validation in performance evaluation. Since dMRI retrieval is more challenging than the retrieval of MRI and PET due to its high dimensionality and complicated processing pipeline, such as artifact correction, tensor estimation, tractography reconstruction and template mapping, we therefore used EN to select the features to further enhance the performance. Notice that the testing data in each trial of leave-one-out was excluded in feature selection, and only NC and AD subjects were used in EN, since there were large overlaps between MCI and the other two groups. We used grid-search to obtain the parameters in PGF, including the size of the neighborhood k and the propagation parameter α. As in previous experiments, MAP was used as the metric to evaluate the performance.

Table 7.5 summarizes the results on individual groups and the overall results. With the multiple types of features, including both the regional mean statistics and the fiber tracking-based features, we compared the proposed PGFg to the naive vector concatenation (CONCAT) method. The experiments with and without EN feature selection were also compared. The PGFg algorithm with features filtered by EN (EN-PGFg) outperformed all the other combinations with the highest MAP of 66.46%. EN-PGFg also showed an even performance on individual groups compared to the EN-CONCAT method. However, PGF did not outperform the simple concatenation (CONCAT) without EN filtering. This fact implied that feature selection is a key step to filter out the noisy and biased patterns falsely revealed by the dMRI features, thus enable PGF to construct more reliable local neighborhoods.

7.2.5   PGF in Multimodal Classification

PGF is inherently an unsupervised model, which can be straightforwardly used in content-based retrieval. However, it also has an application in the multimodal classification. In other words, we can use the equilibrium distribution derived from affinity matrix fusion as a high-level feature representation in classification. To test this idea, we carried out two classification experiments on the ADNI MRI subset, i.e., (1) distinguishing AD (positive) from NC (negative), and (2) distinguishing MCI (positive) from NC (negative). We used the equilibrium distributions obtained by PGFg

Table 7.6 The NC versus AD classification performance (%) of PGFg-SVM and MK-SVM, evaluated on the ADNI MRI subset. Table reproduced with permission from [26]

NC versus AD	Accuracy	Sensitivity	Specificity
MK-SVM	76.5 ± 0.9	80.2 ± 1.5	72.2 ± 1.0
PGFg-SVM	73.7 ± 7.7	69.4 ± 15.5	77.4 ± 11.1

Table 7.7 The NC versus MCI classification performance (%) of PGFg-SVM and MK-SVM, evaluated on the ADNI MRI subset. Table reproduced with permission from [26]

NC versus MCI	Accuracy	Sensitivity	Specificity
MK-SVM	65.6 ± 1.0	37.5 ± 2.1	80.9 ± 1.1
PGFg-SVM	64.9 ± 4.2	75.7 ± 4.0	45.0 ± 10.5

to train a SVM classifier, and compared our PGFg-SVM method to the MK-SVM method [34] using 10-fold cross-validation. EN was again applied to select the features for MK-SVM, and the parameters and hyper-parameters for both PGFg-SVM and MK-SVM were optimized by random search [1].

Tables 7.6 and 7.7 show that PGFg-SVM matched the performance of MK-SVM in both experiments. PGFg-SVM had a similar classification accuracy compared to that of MK-SVM, but there was a dramatic difference between the set two methods in sensitivity and specificity. These findings implied that MK-SVM and PGFg-SVM had different strengths in multimodal classification, and could potentially enhance each other. It was evident that the proposed PGF algorithm was more robust to the large inter-subject variances, whereas the MK method could perform better with small inter-subject variance. PGFg-SVM also had greater standard deviations than MK-SVM, for PGF depended heavily on relationships between subjects. In this analysis, the classification performance of PGF might be restricted by the relatively small training set, but it has a great potential to be improved when big databases become available.

7.3 Summary

In this chapter, we presented a propagation graph fusion method for subject-centered multimodal medical content-based retrieval. Multimodal data could complement the deficiencies of single-modal neuroimaging data in retrieval, as suggested by the results of our experiments. In addition, the proposed PGF enables adaptively reshaping the connections between subjects in the different feature spaces, which led to more reliable retried subjects. PGF is a non-parametric unsupervised model that requires no prior knowledge, and could reduce the variations in both features and the queries. Therefore, it outperformed all of the other state-of-the-art retrieval methods in our experiments, and showed a great potential in multimodal neuroimaging data management.

The PGF method can be further extended to neuroimaging classification. PGF with geometric mean achieved comparable performance with the MK-SVM method in classification. Theoretically, PGF is more robust to the large inter-subject variations compared to the MK-SVM method, but further investigation are needed to confirm this with additional experiments on larger datasets. PGF was also tested in content-based retrieval of brain dMRI data to retrieve subjects based on the their affinity to the subjects. Different dMRI features were derived and selected from the dMRI data, and the PGF algorithm successfully fused the neighborhoods obtained from different feature spaces. Compared to other possible configurations of this proposed dMRI retrieval framework, when PGF was combined with Elastic Net, it achieved the best performance in retrieving the patients diagnosed with AD. The results indicated that dMRI features was potentially a good candidate to index dMRI databases for content-based retrieval, and offer the opportunities to carry out research on the more precisely targeted subjects.

References

1. Bergstra, J., & Bengio, Y. (2012). Random search for hyper-parameter optimization. *Journal of Machine Learning Research, 13*, 281–305. ISSN: 1532-4435.
2. Buckner, R. L., Koutstaal, W., Schacter, D. L., Wagner, A. D., & Rosen, B. R. (1998). Functional-anatomic study of episodic retrieval using fMRI: I. Retrieval effort versus retrieval success. *NeuroImage, 7*, 151–162. ISSN: 1053-8119.
3. Cai, W., Feng, D., & Fulton, R. (2000). Content-based retrieval of dynamic PET functional images. *IEEE Transactions on Information Technology in Biomedicine, 4*, 152–158.
4. Cai, W., Kim, J., & Feng, D. (2008). Content-based medical image retrieval. In D. Feng (Ed.), *Biomedical information technology* (pp. 83–113). Melbourne: Elsevier.
5. Cai, W., et al. (2010). 3D neurological image retrieval with localized pathology-centric CMR-Glc patterns. In *The 17th IEEE international conference on image processing (ICIP)* (3201–3204). IEEE.
6. Cai, W., et al. (2014) Automated feedback extraction for medical imaging retrieval. In *IEEE international symposium on biomedical imaging: From nano to macro (ISBI)* (907–910). IEEE.
7. Che, H., et al. (2014) Co-neighbor multi-view spectral embedding for medical contentbased retrieval. In *IEEE international symposium on biomedical imaging: From nano to macro (ISBI)* (911–914). IEEE.
8. Foncubierta-Rodríguez, A., Depeursinge, A. & Müller, H. (2011). Using multiscale visual words for lung texture classification and retrieval. In H. Müller, H. Greenspan & T. Syeda-Mahmood (Eds.), *The MICCAI workshop on medical content-based retrieval for clinical decision support* (Vol. 7075, pp. 69–79). Berlin, Heidelberg: Springer. ISBN: 978-3-642-28459-5.
9. Liu, S., et al. (2010). A robust volumetric feature extraction approach for 3D neuroimaging retrieval. In *The 32nd annual international conference of the IEEE engineering in medicine and biology society (EMBC)* (5657–5660). IEEE.
10. Liu, S., et al. (2010). Localized multiscale texture based retrieval of neurological image. In *The 23rd IEEE international symposium on computer-based medical systems (CBMS)* (243–248). IEEE.
11. Liu, S., Cai, W., Wen, L., & Feng, D. (2011). Volumetric congruent local binary patterns for 3D neurological image retrieval. In P. Delmas, B. Wuensche, & J. James (Eds.), *The 26th*

international conference on image and vision computing New Zealand (IVCNZ) (272–276). IVCNZ.

12. Liu, S., et al. (2011). Generalized regional disorder-sensitive-weighting scheme for 3D neuroimaging retrieval. In *The 33rd annual international conference of the IEEE engineering in medicine and biology society (EMBC)* (7009–7012). IEEE.

13. Liu, S., et al. (2011). Localized functional neuroimaging retrieval using 3d discrete curvelet transform. In *IEEE international symposium on biomedical imaging: From nano to macro (ISBI)* (1877–1880). IEEE.

14. Liu, S., Cai, W., Wen, L. & Feng, D. (2012). Multiscale and multiorientation feature extraction with degenerative patterns for 3d neuroimaging retrieval. In *The 19th IEEE international conference on image processing (ICIP)* (1249–1252). IEEE.

15. Liu, S., Cai, W., Wen, L., & Feng, D. (2012). Semantic-word-based image retrieval for neurodegenerative disorders. *Journal of Nuclear Medicine, 53*, 2309.

16. Liu, S., Cai, W., Wen, L. & Feng, D. (2013). Multi-channel brain atrophy pattern analysis in neuroimaging retrieval. In *IEEE international symposium on biomedical imaging: From nano to macro (ISBI)* (206–209). IEEE.

17. Liu, S., Cai, W., Wen, L. & Feng, D. (2013). Neuroimaging biomarker based prediction of Alzheimer's disease severity with optimized graph construction. In *IEEE international symposium on biomedical imaging: From nano to macro (ISBI)* (1324–1327). IEEE.

18. Liu, S., et al. (2013). A bag of semantic words model for medical content-based retrieval. In T. Syeda-Mohmood, H. Greenspan, & A. Madahushi (Eds.), *The MICCAI workshop on medical content-based retrieval for clinical decision support (MCBR-CDS)* (1–8). IBM Press.

19. Liu, S., et al. (2013). A supervised multiview spectral embedding method for neuroimaging classification. In *The 20th IEEE international conference on image processing (ICIP)* (601–605). IEEE.

20. Liu, S. Q., et al. (2014). High-level feature based PET image retrieval with deep learning architecture. *Journal of Nuclear Medicine, 55*, 2018.

21. Liu, S., Liu, S. Q., Pujol, S., Kikinis, R. & Cai, W. (2014) Propagation graph fusion for multimodal medical content-based retrieval. In *The 13th annual international conference on control, automation, robotics and vision (ICARCV)* (849–854). IEEE.

22. Liu, S., et al. (2014). Multi-channel neurodegenerative pattern analysis and its application in Alzheimer's disease characterization. *Computerized Medical Imaging and Graphics, 38*, 436–444. ISSN: 0895-6111.

23. Liu, S. Q., et al. (2015). Content-based retrieval of brain diffusion magnetic resonance image. In *Multimodal retrieval in the medical domain* (Vol. 9059). Springer

24. Liu, S. Q., et al. (2015). Longitudinal brain MR retrieval with diffeomorphic demons registration: What happened to those patients with similar changes? In *IEEE international symposium on biomedical imaging: From nano to macro (ISBI)* (588–591). IEEE.

25. Liu, S. Q., et al. (2015). Multi-modal neuroimaging feature learning for multi-class diagnosis of Alzheimer's disease. *IEEE Transactions on Biomedical Engineering, 62*, 1132–1140.

26. Liu, S., et al. (2015). Subject-centered multi-view neuroimaging analysis. In *The 22nd IEEE international conference on image processing (ICIP)*. IEEE.

27. Müller, H., Michoux, N., Bandon, D., & Geissbuhler, A. (2004). A review of content-based image retrieval systems in medical applications-clinical benefits and future directions. *International Journal of Medical Informatics, 73*, 1–23.

28. Park, H. (2012). ISOMAP induced manifold embedding and its application to Alzheimer's disease and mild cognitive impairment. *Neuroscience Letters, 513*, 141–145. ISSN: 0304-3940.

29. Ramírez, J., et al. (2009). Early detection of the Alzheimer disease combining feature selection and kernel machines. In M. Köppen, N. Kasabov, & G. Coghill (Eds.), *Advances in neuroinformation processing* (Vol. 5507, pp. 410–417). Berlin, Heidelberg: Springer. ISBN: 978-3-642-03039-0.

30. Shen, L., et al. (2011). Identifying neuroimaging and proteomic biomarkers for MCI and AD via the elastic net. In T. Liu, D. Shen, L. Ibanez, & X. Tao (Eds.), *Multimodal brain image analysis (MBIA)* (Vol. 7012, pp. 27–34). Berlin, Heidelberg: Springer. ISBN: 978-3-642-24445-2.

31. Shen, H., Tao, D., & Ma, D. (2013). Multiview locally linear embedding for effective medical image retrieval. *PLoS ONE*, *8*, e82409.
32. Unay, D., Ekin, A., & Jasinschi, R. (2010). Local structure-based region-of-interest retrieval in brain MR images. *IEEE Transactions on Information Technology in Biomedicine*, *14*, 897–903.
33. Xia, T., Tao, D., Mei, T., & Zhang, Y. (2010). Multiview spectral embedding. *IEEE Transactions on Systems, Man, and Cybernetics, Part B: Cybernetics*, *40*, 1438–1446.
34. Zhang, D., Wang, Y., Zhou, L., Yuan, H., & Shen, D. (2011). Multimodal classification of Alzheimer's disease and mild cognitive impairment. *NeuroImage*, *55*, 856–867. ISSN: 1053-8119.
35. Zhang, S., Yang, M., Cour, T., Yu, K. & Metaxas, D. (2012). Query specific fusion for image retrieval English. In A. Fitzgibbon, S. Lazebnik, P. Perona, Y. Sato & C. Schmid (Eds.), *European conference on computer vision (ECCV)* (660–673). Berlin, Heidelberg: Springer. ISBN: 978-3-642-33708-6.
36. Zhang, L., et al. (2013). Graph cuts based relevance feedback in image retrieval. In *The 20th IEEE international conference on image processing (ICIP)* (4358–4362). IEEE.
37. Zhang, F., et al. (2014). Latent semantic association analysis for medical image retrieval. In *International conference on digital image computing: Techniques and applications (DICTA)* (pp. 1–6).
38. Zhang, F., et al. (2015). Dictionary refinement with visual word significance for medical image retrieval. *Neurocomputing* (2015).
39. Zhang, F., et al. (2015). Ranking-based vocabulary pruning in bag-of-features for image retrieval. In *The 1st Australian conference on artificial life and computational intelligence (ACALCI)* (Vol. 8955, pp. 436–445). Springer.

Chapter 8
Conclusions and Future Directions

A series of models and methods have been developed to systematically analyze the neurodegeneration from data acquisition to application development. This chapter concludes the research findings in neurodegenerative disorder based on the analysis on large-scale multimodal datasets and further outlines the future directions.[1]

8.1 Conclusions

In the data computing layer, four ADNI subsets, i.e., MRI, PET, dMRI and MRI-PET, consisting of over 800 subjects were included in this study. Different image computing protocols, i.e., artifact correction, spatial/functional normalization, registration, segmentation, labeling and parameter estimation, were used for these subsets to ensure the quality of the datasets.

In the feature representation layer, a set of hand-engineered morphological features, such as convexity (CNV) and solidity (SLD) [14, 15] and functional features, such as Difference-of-Gaussian Mean (DoG-M), Contrast (DoG-C) and Z-score (DoG-Z) [2, 3] were proposed to model changes caused by neurodegeneration, i.e., atrophy and hypo-metabolism. These features were compared with other state-of-the-art features, including Grey Matter Volume (GMV), Local Gyrification Index (LGI), Mean Index (M-IDX), and Fuzzy Index (F-IDX) [25]. Each feature has a particular strength and no single feature could completely outperform the other features. In the experiment of classifying Alzheimer's disease (AD), mild cognitive impairment (MCI) and normal control (NC) subjects, F-IDX had the highest overall accuracy and specificity, whereas DoG-M had the highest sensitivity, and grey matter volume achieved higher precision in detecting AD.

---

[1]Some content of this chapter has been reproduced with permission from [22, 24].

© Springer Nature Singapore Pte Ltd. 2017
S. Liu, *Multimodal Neuroimaging Computing for the Characterization*
of Neurodegenerative Disorders, Springer Theses, DOI 10.1007/978-981-10-3533-3_8

Given the multimodal inputs, the proposed stacked auto-encoder [21], which was trained using a 'zero masking' strategy and fine-tuned with a softmax layer, out-performed many state-of-the-art methods, such as support vector machine (SVM), multiple kernel SVM (MK-SVM), and stacked auto-encoders with multiple kernel SVM (SAE-MK-SVM). It is more powerful in capturing the complex correlations between brain regions, features, and modalities, as demonstrated in the most challenging experiment, i.e. classifying the MCI converter (cMCI) and MCI non-converters (ncMCI).

In the pattern analysis layer, the channel-based and the view-based analysis methods were used to investigate the patterns of disease pathologies derived with different analysis methods, such as t-test, SVM and elastic net (EN) [12, 18], or different features, such as GMV, LGI, M-IDX, F-IDX and our proposed features [25]. Based on the results of multi-channel analysis, a set of key brain regions were found to be associated with neurodegeneration, including the frequently reported hippocampus, cingulate gyrus, temporal lobe and parietal lobe, as well as some new brain ROIs, including temporal pole, brainstem, subgenual and presuggenual frontal cortex. Further exploration using the cross-view analysis method showed the features can be grouped to different clusters according to their patterns, and the highest classification rates were always achieved by the combination of features from different sub-clusters, e.g., GMV/DoG-C, GMV/F-IDX, or GMV/F-IDX.

In the application layer, a range of clinical applications were developed to translate the clinical findings to improved diagnostic or management tools. One application is to enhance the staging of AD by integrating the domain knowledge to the graph-cut algorithm [11]. A new global-cost function was designed to modify the objective function of graph-cut algorithm, and this optimized algorithm significantly improve the distinction between patient groups. In another application, a novel multifold Bayesian kernelization (MBK) was developed for multimodal multi-class classification of AD [16]. MBK is independent to the classifiers and infers the patient's diagnosis by synthesizing the output prediction probabilities of multimodal biomarkers. The prediction can be made based on arbitrary number of biomarkers in MBK, and the performance gain is transferable to other classifiers as references for feature selection.

Furthermore, a propagation graph fusion (PGF) algorithm [17] developed to provide clinical decision support via subject-centered content-based neuroimaging retrieval. As an unsupervised method, PGF does not require any prior knowledge of the features or the queries, but could adaptively reshape the connections between the subjects according to query, and is able to find more relevant subjects. The original PGF algorithm was further improved by enforcing the conformity of the retrieval results across multiple views using the geometric-mean-based fusion method [23]. All of these applications show great potential in the management of AD and MCI.

8.2   Future Directions

Neuroimaging technologies are still on the fast track for development towards higher spatial, temporal and angular resolutions, shorter acquisition time, stronger magnetic fields, and lower scanning cost. In particular, advances in the hybrid imaging scanners, e.g., PET/CT and PET/MRI, will become available in more clinics and laboratories, enabling more interesting discoveries in the neuropsychiatric disorders [22, 24].

The improved imaging capabilities will offer more opportunities to identify the effective biomarkers for distinguishing the neuropsychiatric disorders, or various pathologies of the same disorder with higher statistical power [1, 4–10]. These biomarkers will validated through large-scale evaluation in existing clinical laboratories and standardized in clinical trials. Once the biomarkers reach a satisfactory level or the treatment, appropriate clinical guidelines should be developed to support and encourage widespread clinical testing.

Another direction of neuroimaging computing will lie in the integration of imaging analyses with the other imaging or non-imaging analyses. One example on dMRI [19] is given in Chap. 7, Sect. 7.2.4, and another example on multimodal retrieval is given in our previous study [13]. Neuroimaging computing, which is inherently an interdisciplinary research area, will keep attracting neurology and psychiatry practitioners, imaging specialists, computer scientists, engineers, physicists, and other researchers who are interested in the development of innovative neurotechnologies and translational applications. Imaging genetics, for example, is one of the most promising areas with potential significant advances, which aims to discover the genetic basis of brain anatomical and functional abnormalities and their connections with the neurological and psychiatric disorders. It is trending to use the image-based biomarkers for brain disorders to reveal the endophenotypes for various gene mutations. The combination of image-based and genetic biomarkers will help us better understand the physiopathology, and enhance the early diagnosis and management of these brain disorders.

Neuroimaging data will continue to grow in complexity and volume, and demand improvements of the neuroimaging computing models and methods. The importance of longitudinal information, for instance, has been highly recognized and should be integrated into the analysis pipeline whenever is possible. One example on longitudinal analysis [20] is given in Chap. 2, Sect. 2.4.2. As large-scale longitudinal datasets are being collected, they will eventually enable us to understand the physiopathological processes of the disorders and approximate their progression trajectories, which will have many translational impacts in early diagnosis and prognosis.

Future studies will also continue to tilt towards the subject-centered analyses, since there will always be inter-subject variations in a database, however large and comprehensive. Two examples of subject-centered analysis methods, i.e., PGF [17] and its improved version [23], are discussed in detail in Chap. 7. Therefore, innovative approaches for personalized/subject-centered analyses are highly demanded, and will be the most important part of translational neuroimaging research.

In addition, the neuroimaging computing models and methods will continue increase the grade of accuracy, reproducibility, robustness, semantic interpretation, automation, user interaction, and eventually standardized in a form that is readily to be integrated into the clinical workflows and facilitate clinical testing of the new neuroimaging biomarkers [3, 5, 9, 13, 26–29].

References

1. Cai, W., et al. (2010). 3D neurological image retrieval with localized pathology-centric CMR-Glc patterns. In *The 17th IEEE International Conference on Image Processing (ICIP)* (pp. 3201–3204). IEEE.
2. Cai, W., et al. (2014). A 3D difference of gaussian based lesion detector for brain PET. In *IEEE International Symposium on Biomedical Imaging: From Nano to Macro (ISBI)* (pp. 677–680). IEEE.
3. Cai, W., et al. (2014). Automated feedback extraction for medical imaging retrieval. In *IEEE International Symposium on Biomedical Imaging: From Nano to Macro (ISBI)* (pp. 907–910). IEEE.
4. Liu, S., et al. (2010). Localized multiscale texture based retrieval of neurological image. In *The 23rd IEEE International Symposium on Computer-Based Medical Systems (CBMS)* (pp. 243–248). IEEE.
5. Liu, S., et al. (2010). A robust volumetric feature extraction approach for 3D neuroimaging retrieval. In *The 32nd Annual International Conference of the IEEE Engineering in Medicine and Biology Society (EMBC)* (pp. 5657–5660). IEEE.
6. Liu, S., Cai, W., Wen, L., & Feng, D. (2011). Volumetric congruent local binary patterns for 3D neurological image retrieval. In P. Delmas, B. Wuensche, & J. James (Eds.), *The 26th International Conference on Image and Vision Computing New Zealand (IVCNZ)* (pp. 272–276). IVCNZ.
7. Liu, S., et al. (2011). Generalized regional disorder-sensitive-weighting scheme for 3D neuroimaging retrieval. In *The 33rd Annual International Conference of the IEEE Engineering in Medicine and Biology Society (EMBC)* (pp. 7009–7012). IEEE.
8. Liu, S., et al. (2011). Localized functional neuroimaging retrieval using 3D discrete curvelet transform. In *IEEE International Symposium on Biomedical Imaging: From Nano to Macro (ISBI)* (pp. 1877–1880). IEEE.
9. Liu, S., Cai, W., Wen, L., & Feng, D. (2012). Semantic-word-based image retrieval for neurodegenerative disorders. *Journal of Nuclear Medicine, 53*, 2309.
10. Liu, S., Cai, W., Wen, L., & Feng, D. (2012). Multiscale and multiorientation feature extraction with degenerative patterns for 3D neuroimaging retrieval. In *The 19th IEEE International Conference on Image Processing (ICIP)* (pp. 1249–1252). IEEE.
11. Liu, S., Cai, W., Wen, L., & Feng, D. (2013). Neuroimaging biomarker based prediction of Alzheimer's disease severity with optimized graph construction. In *IEEE International Symposium on Biomedical Imaging: From Nano to Macro (ISBI)* (pp. 1324–1327). IEEE.
12. Liu, S., Cai, W., Wen, L., & Feng, D. (2013). Multi-channel brain atrophy pattern analysis in neuroimaging retrieval. In *IEEE International Symposium on Biomedical Imaging: From Nano to Macro (ISBI)* (pp. 206–209). IEEE.
13. Liu, S., et al. (2013). A bag of semantic words model for medical content-based retrieval. In T. Syeda-Mohmood, H. Greenspan, & A. Madahushi (Eds.), *The MICCAI Workshop on Medical Content-based Retrieval for Clinical Decision Support (MCBR-CDS)* (pp. 1–8). IBM Press.
14. Liu, S., et al. (2013). A supervised multiview spectral embedding method for neuroimaging classification. In *The 20th IEEE International Conference on Image Processing (ICIP)* (pp. 601–605). IEEE.

15. Liu, S., et al. (2013). Localized sparse code gradient in Alzheimer's disease staging. In *The 35th Annual International Conference of the IEEE Engineering in Medicine and Biology Society (EMBC)* (pp. 5398–5401). IEEE.

16. Liu, S., et al. (2013). Multifold Bayesian kernelization in Alzheimer's diagnosis. In K. Mori, I. Sakuma, Y. Sato, C. Barillot, & N. Navab (Eds.), *The 16th International Conference on Medical Image Computing and Computer-Assisted Intervention (MICCAI)* (Vol. 8150, pp. 303–310). Berlin: Springer.

17. Liu, S., Liu, S. Q., Pujol, S., Kikinis, R., & Cai, W. (2014). Propagation graph fusion for multi-modal medical content-based retrieval. In *The 13th Annual International Conference on Control, Automation, Robotics and Vision (ICARCV)* (pp. 849–854). IEEE.

18. Liu, S., et al. (2014). Multi-channel neurodegenerative pattern analysis and its application in Alzheimer's disease characterization. *Computerized Medical Imaging and Graphics, 38,* 436–444. ISSN: 0895-6111.

19. Liu, S. Q., et al. (2015). Content-based retrieval of brain diffusion magnetic resonance image. *Multimodal retrieval in the medical domain* (Vol. 9059). Berlin: Springer.

20. Liu, S. Q., et al. (2015). Longitudinal brain MR retrieval with diffeomorphic demons registration: What happened to those patients with similar changes? In *IEEE International Symposium on Biomedical Imaging: From Nano to Macro (ISBI)* (pp. 588–591). a: IEEE.

21. Liu, S. Q., et al. (2015). Multi-modal neuroimaging feature learning for multi-class diagnosis of Alzheimer's disease. *IEEE Transactions on Biomedical Engineering, 62,* 1132–1140.

22. Liu, S., et al. (2015). Multimodal neuroimaging computing: A review of the applications in neuropsychiatric disorders. *Brain Informatics, 2,* 167–180.

23. Liu, S., et al. (2015). Subject-centered multi-view neuroimaging analysis. In *The 22nd IEEE International Conference on Image Processing (ICIP).* IEEE.

24. Liu, S., et al. (2015). Multimodal neuroimaging computing: The workflows. *Methods and Platforms. Brain Informatics, 2,* 181–195.

25. Liu, S., et al. (2016). Cross-view neuroimage pattern analysis for Alzheimer's disease staging. *Frontiers in Aging Neuroscience.*

26. Zhang, L., et al. (2013). Graph cuts based relevance feedback in image retrieval. In *The 20th IEEE International Conference on Image Processing (ICIP).* IEEE.

27. Zhang, F., et al. (2014). Latent semantic association analysis for medical image retrieval. In *International Conference on Digital Image Computing: Techniques and Applications (DICTA)* (pp. 1–6).

28. Zhang, F., et al. (2014). Semantic association for neuroimaging classification of PET images. *Journal of Nuclear Medicine, 55,* 2029.

29. Zhang, F., et al. (2015). Pairwise latent semantic asociation for similarity computation in medical imaging. *IEEE Transactions on Biomedical Engineering.*

Appendix A: Abbreviations and Acronyms

Disorders and Pathologies

AD	Alzheimer's Disease
ADHD	Attention-Deficit Hyperactivity Disorder
ASD	Autism Spectrum Disorder
CBD	Corticobasal Degeneration
CJD	Jacob-Creutzfeldt Disease
CSF	Cerebrospinal Fluid
DALYs	Disability-Adjusted Life-Years
DLBD	Dementia with Lewy Bodies
FTD	Frontotemporal Dementia
MCI	Mild Cognitive Impairment
MND	Motor Neuron Disease
MS	Multiple Sclerosis
MSA	Multiple System Atrophy
NC	Normal Control
NFT	Neurofibrillary Tangle
OCD	Obsessive‚ÄìCompulsive Disorder
PD	Parkinson's Disease
PSP	Progressive Supranuclear Palsy
RSN	Resting State Network
VD	Vascular Dementia

Imaging Modalities and Measures

ACI	Amyloid Convergence Index
AXD	Axial Diffusivity
BCI	Brain Computer Interface
CMRGlc	Cerebral Metabolic Rate of Glucose
CNV	Convexity Ratio
CSD	Constrained Spherical Deconvolution
CT	Computed Tomography

© Springer Nature Singapore Pte Ltd. 2017
S. Liu, *Multimodal Neuroimaging Computing for the Characterization of Neurodegenerative Disorders*, Springer Theses, DOI 10.1007/978-981-10-3533-3

DCVT	Discrete Curvelet Transform
DoG-C	Difference-of-Gaussian Contrast
DoG-M	Difference-of-Gaussian Mean
DoG-Z	Difference-of-Gaussian Z-score
DSI	Diffusion Spectral Imaging
DTI	Diffusion Tensor Imaging
ERF	Event-Related Fields
ERP	Event-Related Potentials
FA	Fractional Anisotropy
FDG	2-deoxy-2-$[^{18}$F]fluoro-D-glucose
FRACT	Q-Ball Imaging with Funk-Radon and Cosine Transform
F-IDX	Fuzzy Index
GLCM	Grey Level Co-occurrence Matrix
GMD	Grey Matter Density
GMV	Grey Matter Volume
GQI	Generalized Q-sampling Imaging
HCI	Hypo-metabolic Convergence Index
LBP	Local Binary Pattern
LGI	Local Gyrification Index
MD	Mean Diffusivity
MRI	Magnetic Resonance Imaging
M-IDX	Mean Index
PET	Positron Emission Tomography
QBI	Q-Ball Imaging
RD	Radial Diffusivity
SLD	Solidity Ratio
SPECT	Single Photon Emission Computed Tomography
SUVR	Standard Uptake Value Ratios

Research Organizations and Administrations

AADRF	Alzheimer's Australia Dementia Research Foundation
ADNI	Alzheimer's Disease Neuroimaging Initiatives
AIBL	Australian Imaging Biomarker and Lifestyle Flagship Study of Aging
BRAIN	Brain Research through Advancing Innovative Neurotechnologies Initiative
BMIT	Biomedical and Multimedia Information Technology Research Group
FDA	Food and Drug Administration
HBP	Human Brain Project
HCP	Human Connectome Project
ICBM	International Consortium for Brain Mapping
MNI	Montreal Neurological Institute
NIA	National Institute of Aging
NIBIB	National Institute of Biomedical Imaging and Bioengineering
OECD	Organization for Economic Co-operation and Development

RPAH	Royal Prince Alfred Hospital
SPL	Surgical Planning Laboratory
SUGUNA	Sydney University Graduate Union North America

Algorithms and Applications

ANOVA	Analysis of Variance
BoW	Bag of Words Model
CAD	Computer Aided Diagnosis
CBIR	Content-based Image Retrieval
DOM	Disorder-Oriented Mask
EN	Elastic Net
GC	Graph Cut
IGT	Image-Guided Therapy
MAP	Mean Average Precision
MAPER	Multi-Atlas Propagation with Enhanced Registration
MBK	Multifold Bayesian Kernelization
MSE	Multi-view Spectral Embedding
PCA	Principle Component Analysis
PGF	Propagation Graph Fusion
SAE	Sparse Auto-Encoder
SVM	Support Vector Machine
TOST	Two One-Sided Test

Appendix B: ICBM Brain Template

The full list of 83 brain regions defined in the ICBM_152 brain template. Left column: the region indexes. Right column: the region labels.

Region Index	Region Label
1	Hippocampus (Right)
2	Hippocampus (Left)
3	Amygdala (Right)
4	Amygdala (Left)
5	Anterior temporal lobe (Right)
6	Anterior temporal lobe (Left)
7	Anterior temporal lobe, lateral part (Right)
8	Anterior temporal lobe, lateral part (Left)
9	Parahippocampal and ambient gyri (Right)
10	Parahippocampal and ambient gyri (Left)
11	Superior temporal gyrus, posterior part (Right)
12	Superior temporal gyrus, posterior part (Left)
13	Middle and inferior temporal gyrus (Right)
14	Middle and inferior temporal gyrus (Left)
15	Fusiform gyrus (Right)
16	Fusiform gyrus (Left)
17	Cerebellum (Right)
18	Cerebellum (Left)
19	Brainstem (unpaired)
20	Insula (Right)
21	Insula (Left)
22	Lateral remainder of occipital lobe (Right)
23	Lateral remainder of occipital lobe (Left)
24	Cingulate gyrus, anterior part (Right)
25	Cingulate gyrus, anterior part (Left)
26	Cingulate gyrus, posterior part (Right)
27	Cingulate gyrus, posterior part (Left)
28	Middle frontal gyrus (Right)

(continued)

© Springer Nature Singapore Pte Ltd. 2017
S. Liu, *Multimodal Neuroimaging Computing for the Characterization of Neurodegenerative Disorders*, Springer Theses, DOI 10.1007/978-981-10-3533-3

(continued)

Region Index	Region Label
29	Middle frontal gyrus (Left)
30	Posterior temporal lobe (Right)
31	Posterior temporal lobe (Left)
32	Inferiolateral remainder of parietal lobe (Right)
33	Inferiolateral remainder of parietal lobe (Left)
34	Caudate nucleus (Right)
35	Caudate nucleus (Left)
36	Nucleus accumbens (Right)
37	Nucleus accumbens (Left)
38	Putamen (Right)
39	Putamen (Left)
40	Thalamus (Right)
41	Thalamus (Left)
42	Pallidum (Right)
43	Pallidum (Left)
44	Corpus callosum (unpaired)
45	Lateral ventricle - (apart temporal horn) (Right)
46	Lateral ventricle - (apart temporal horn) (Left)
47	Lateral ventricle, temporal horn (Right)
48	Lateral ventricle, temporal horn (Left)
49	Third ventricle (unpaired)
50	Precentral gyrus (Right)
51	Precentral gyrus (Left)
52	Straight gyrus (Right)
53	Straight gyrus (Left)
54	Anterior orbital gyrus (Right)
55	Anterior orbital gyrus (Left)
56	Inferior frontal gyrus (Right)
57	Inferior frontal gyrus (Left)
58	Superior frontal gyrus (Right)
59	Superior frontal gyrus (Left)
60	Postcentral gyrus (Right)
61	Postcentral gyrus (Left)
62	Superior parietal gyrus (Right)
63	Superior parietal gyrus (Left)
64	Lingual gyrus (Right)
65	Lingual gyrus (Left)
66	Cuneus (Right)
67	Cuneus (Left)
68	Medial orbital gyrus (Right)
69	Medial orbital gyrus (Left)
70	Lateral orbital gyrus (Right)
71	Lateral orbital gyrus (Left)
72	Posterior orbital gyrus (Right)
73	Posterior orbital gyrus (Left)
74	Substantia nigra (Right)
75	Substantia nigra (Left)
76	Subgenual frontal cortex (Right)
77	Subgenual frontal cortex (Left)
78	Subcallosal area (Right)
79	Subcallosal area (Left)
80	Pre-subgenual frontal cortex (Right)
81	Pre-subgenual frontal cortex (Left)
82	Superior temporal gyrus, anterior part (Right)
83	Superior temporal gyrus, anterior part (Left)

Printed in the United States
By Bookmasters